Darwin
and Darwinism

PROBLEMS IN EUROPEAN CIVILIZATION

Under the editorial direction of
John Ratté
Amherst College

Darwin and Darwinism

Revolutionary Insights concerning Man, Nature, Religion, and Society

Edited and with an introduction by

Harold Y. Vanderpool
Wellesley College

D. C. HEATH AND COMPANY
Lexington, Massachusetts   Toronto   London

Copyright © 1973 by D. C. Heath and Company

All rights reserved. No part of this publication may be reproduced or transmitted in any form or by any means, electronic or mechanical, including photocopy, recording, or any information storage or retrieval system, without permission in writing from the publisher.

Published simultaneously in Canada.

Printed in the United States of America.

International Standard Book Number: 0-669-85407-7

Library of Congress Catalog Card Number: 73-7052

CONTENTS

INTRODUCTION — vii
THE CLASH OF ISSUES — xxxii

I BEFORE THE ORIGIN OF SPECIES (1859)

Genesis 1:1–2:9, 2:15–25
THE BIBLICAL STORY OF CREATION — 3

J. J. S. Perowne
THE CHARACTER OF RELIGION — 7

James D. Dana
THE CHARACTER OF SCIENCE — 12

William Paley
PROOF OF GOD FROM THE WORLD OF BIOLOGY — 18

William Cowper
HYMNS TO THE GOD OF NATURE — 27

William Wordsworth
NATURE AS MORAL GUIDE AND DIVINE PRESENCE — 30

Psalms 8:1, 3–9; Romans 1:18–20, 28–32
MANKIND'S PROXIMITY TO THE ANGELS — 34

William Kirby
THE NOBLEST OF SPECIES — 36

II EVOLUTION, NATURE AND RELIGION

Charles Darwin
A NEW, REVOLUTIONARY WORLD VIEW — 43

Samuel Wilberforce and Adam Sedgwick
TRADITIONAL RELIGION AND SCIENCE OPPOSE EVOLUTION — 75

T. H. Huxley
ORTHODOXY SCOTCHED, IF NOT SLAIN   91

Leslie Stephen
SPREADING AGNOSTICISM   104

Asa Gray
THE COMPATIBILITY OF EVOLUTION AND RELIGION   113

Baden Powell
THE VALIDATION OF RELIGION APART FROM RATIONAL PROOF   123

S. R. Driver
A NEW UNDERSTANDING OF THE BIBLE   133

Thomas Hardy and Alfred Lord Tennyson
THE NEW FACES OF NATURE   139

III   EVOLUTION, MAN AND SOCIETY

Charles Darwin
THE NATURAL EVOLUTION OF MAN AND MORALITY   147

Alfred Russel Wallace
THE INADEQUACIES OF DARWINIAN EVOLUTION   161

John Fiske
THE ASCENT OF MAN   172

T. H. Huxley
SECULAR HUMANISM   181

Thomas Hardy and Algernon Charles Swinburne
MANKIND'S POST-DARWINIAN STATUS: LONELINESS OR LIBERATION?   195

Herbert Spencer
SOCIETY CONDITIONED BY EVOLUTION   199

T. H. Huxley
SOCIETY MODIFIED ACCORDING TO HUMAN STANDARDS   209

SUGGESTIONS FOR ADDITIONAL READING   216

INTRODUCTION

Charles Darwin's theory of evolution created a variety of intense and fascinating reactions. T. H. Huxley became so captivated by its cogency and usefulness that he instantly utilized it as a fundamental perspective in his own controversial and renowned biological research. Huxley spent a lifetime defending Darwinism in the English-speaking world. Karl Marx greeted Darwin's theory as a dramatic confirmation of his naturalistic world view, and in spite of considering Darwin's writing skills deplorable, sought to dedicate *Das Kapital* to him. Darwin so influenced the great English novelist and poet Thomas Hardy that Hardy thoroughly revised his former romantic beliefs. Convinced that romantic images of nature and providence were no longer acceptable, Hardy wrote that sun, rain and forest offered to man neither a providential guide through the mazed confusions of society nor a "soft release" from life's problems. "Crass Casualty obstructs the sun and rain," cried Hardy.[1] No uplifting moral precepts can be drawn from the chance-riddled, brutal struggles of the natural world.

Other thinkers—less known today although widely influential in their own time—were irrevocably influenced by Darwin. Leslie Stephen, a man of letters and the father of Virginia Woolf, simply announced that *The Origin of Species* led him to set aside forever the defunct religious point of view that he tried to believe in as a theological student. And the exceedingly popular and influential social philosopher Herbert Spencer received from Darwin a form of secular confirmation. Spencer believed that Darwin offered scientific support

---

[1] Thomas Hardy, "Hap," in *Collected Poems of Thomas Hardy* (New York: Macmillan, 1926), p. 7.

FIGURE 1. Charles Darwin in 1838, the year in which he discovered the principle of evolution by natural selection. Portrait drawing by T. F. Maguire. *(Radio Times Hulton Picture Library)*

for his laissez-faire social theories, which he envisioned as the necessary and progressive shape of future society.

These reactions indicate both the power and scope of the impact of Darwin's theory of evolution on the Western intellect. Darwin and

Introduction

Darwinists created a veritable revolution that profoundly influenced existing presuppositions about man, religion, the natural world, social institutions, and even the fundamental presupposition that change is a permanent aspect of human life and institutions. Primary writings concerning this Darwinian revolution thus become fascinating and essential reading for our understanding of modern history.

For the most part, however, we have not had ready access to the dramatic documents and debates which, like the effects of a tremendous earthquake on geological terrain, reshaped previous contours of Western thought. This collection initiates an understanding and interpretation of crucial primary sources themselves rather than further reinforcing dependency on secondary interpretations. A succinct, representative reading is often worth five thousand words of interpretation. The readings here are designed to encourage critical evaluation and analysis with respect to the history of ideas. This includes an awareness of some of the ways that Darwinian science influenced literature, religion, philosophy, and social thought.

This collection, however, represents more than a usable and critical device for assessing the nature of Darwin's historical influence. It also encourages a probing and clarifying of present problems. The list of questions raised by Darwin's research forms a kind of flow sheet for many contemporary personal and public concerns. The grounds for human significance—or lack of it, the question as to whether the cosmos is ruled by purpose or chance, the status and content of Western religion, mankind's understanding of and relationship to nature, and the extent to which struggle and competition are legitimated in society were all central issues debated with ultimate seriousness in Darwin's era and no less discussed in our own. These readings are designed to evoke discussion and help clarify issues pertaining to these personal, philosophical, social and religious questions.

Of course, other crucial "intellectual revolutions" dealt with aspects of these same problems and, along with Darwinism, served as formative influences for the making of the "modern mind." Newtonian science before Darwin, and Marxist social thought and Freudian psychology during and after Darwin's era, are notable examples of such revolutionary movements. In this respect, Darwinism was one among other rare and "great illuminations, elucidations, enlightenments" whose powerful beams represent the modern West's laborious and imaginative quest for meaning and knowledge.[2] Like every intellectual revolution, Darwinism represented both the fruition of past research and a catalyst for new perspectives and analysis. The selections here identify and order, from a mass of often confused issues, those questions concerning man, nature, religion and society that are closely or directly related to Darwin's thought.

These dual purposes—historical analysis and a personal grappling with crucial issues—are both served by separating out the issues raised by Darwinism and treating them within a chronological framework. Section I contains representative readings from pre-Darwinian spokesmen. Sections II and III begin with Darwin's own revolutionary perspectives and then move to criticisms, defenses and reformulations of his opinions and of other issues raised by his work. For each issue raised, the reader can assess the way traditional beliefs were destroyed, defended, or reshaped, and ask which point of view is most convincing and why.

The issues and writers in each of these sections warrant attention. The discussions below are divided and related to Sections I through III respectively and are useful both individually and as a whole. Individually, they serve as valuable, separate introductions, to be used as the reader arrives at each new section of readings. As a whole, they give a sense of the crucial significance of specific issues as related to the total picture of change that occurred.

Section I. The young Charles Darwin (1809–1882) grew up in a traditional world exceedingly different from the later one that he played no little part in shaping. Among the fundamental ideas that characterized this traditional world view were the beliefs that all living things were directly created by God; that science and scripture were harmonious; that nature reflected divine ingenuity and providence, providing mankind with an ideal context for self-fulfillment; and that as the noblest of all creatures, man rightly believed that earth, plants, and animals were created for his requirements and comforts.

In its remarkable way, and aside from its age and cultural origins, the first two chapters of the Bible—chosen as the first reading in this section—encased these traditional beliefs and brought them together into an intact, functioning world view. As created by God, the

---

[2] See the conclusion of Stanley Edgar Hyman's *The Tangled Bank* (New York: Atheneum, 1962), p. 447.

Introduction

world of nature—arranged from the lowest organism up to man, the highest of all creatures—was regarded as a changeless "state of order and beauty."[3]

Most educated men and women in England and America before the publication of *The Origin of Species* in 1859 revered the Bible and accepted its teachings as by and large true, including its ideas about the creation of immutable animal and plant species by God. Intellectuals thus commonly assumed that the Bible and science could be harmonized with each other and should inform one another. The young Charles Darwin shared this respect for scripture with his contemporaries. While on board H. M. S. *Beagle,* he quoted the Bible "as an unanswerable authority" to unruly officers on the ship.[4]

This widely held conviction about the truth and accuracy of the Bible was not based on blind, dogmatic attachments to ecclesiastical tradition. Rather, it gave witness to an enormous body of reasoned investigation by members of many Christian denominations. Literal exegesis of the Bible as well as a devotional respect for its accuracy was set in motion by the Reformation and was further heightened by the Enlightenment. Enlightenment Deists had identified the Bible with a superstitious, authoritarian past and believed that they could substantiate theism and essential moral norms by logical arguments independent from scripture. In their fervor to render the past obsolete, Deistic critics sallied forth verbally against the Bible. The orthodox—clearly representing the educated majority after the French and American revolutions—responded with defenses of the literal, historical truth of the Bible. This orthodox legacy of a genuine trust in the accuracy of the Bible still stood in 1859 as a widely assumed intellectual norm. Yet challenges to this norm existed.

In two ways, in fact, trust in the literal accuracy of scripture was being undermined in England and America during the first half of the nineteenth century. First, a selected acceptance of the biblical criticism emanating from German universities was made in England and America by such people as Samuel Taylor Coleridge (1772–1834) and the English and American Unitarians. This new criticism directly challenged traditional opinions about the accuracy of the Bible. But

---

[3] William Latham Bevan, "Earth," in *A Dictionary of the Bible,* ed. by William Smith (Boston: Little, Brown, 1860), p. 463.
[4] Charles Darwin, *The Autobiography of Charles Darwin* (New York: Dover Publications, 1958 [1898], p. 62. For another example, see the references to Darwin's botany professor at Cambridge, the Reverend John Stevens Henslow, in ibid., p. 22.

since it was not widely influential until after 1860, biblical criticism did more to increase the intellectual turmoil in the period after the publication of the *Origin,* than to prepare the stage for Darwin. Second, by 1830 geological investigation regarding the age and development of the earth had undermined the literal accuracy of two stories in the Bible—namely the narratives of the creation of the world and the universal flood under Noah—both from the book of Genesis. These problems led scientific and religious progressives to "deliteralize" the first chapters of Genesis, that is, to interpret the "days" of creation as creative epochs and to regard the flood of Noah as local, not worldwide. Nevertheless, intellectuals regularly maintained that this process did not prove the Bible false nor demonstrate that scripture was incompatible with science.

In short, the great majority of the educated in 1859 trusted the Bible for its accuracy and truth-value in spite of the facts that serious questions had emerged and that some biblical texts had to be interpreted as nonliteral. No public debate had pitted science against scripture, and only lonely individualists were ready to regard fundamental teachings of the Bible like the creation by God of plants, animals and man, and the fixity of species as false. Darwinism radically altered this climate of opinion.

The selections by J. J. S. Perowne (1823–1904) and James D. Dana (1813–1895) represent normal attitudes among religious leaders and scientists respectively, and show how religion and science were harmoniously interconnected at midcentury. As an interpretation of the creation stories in Genesis, Perowne's article indicates how biblical scholars still regarded the Bible as historically trustworthy, yet were beginning to question whether its language had to presuppose the findings of modern science. Perowne knew that geology created problems for a literal understanding of Genesis, yet remained convinced—as did most of his contemporaries—that the Bible and science could be harmonized. Perowne's article also reflects an intelligent clergyman's stance of cautious openness to new information, and displays his refusal to accept easy answers.

England and America abounded with men who thought like Bishop Perowne and William Paley (1743–1805), the orthodox champion of the age. Paley's books were archetypes of lucidity and logic, and had become required reading for the B.A. examinations of students

Introduction

such as Charles Darwin.[5] After considerable uncertainty, Darwin decided to major in classics and religious studies (then called divinity) at Cambridge. His intention to become a clergyman seemed a rather natural alternative for a young man of good social standing who apparently had "no remarkable gifts or inclinations." With the Bible regarded as trustworthy, with religion considered as rationally provable, with opportunities to go hunting and to study nature as a country parson, with greatly respected clergymen as his professors in numerous academic fields, with the encouragement of his father, and with writers like Paley depicting Jesus Christ as a man of "moderation and soundness of judgment," why not join ranks with men of the cloth?[6]

The reading from James Dwight Dana illustrates two significant features about the character of science before 1859. First, it indicates the great extent to which most scientists respected the biblical world view as a source of authority. Their scientific research led them to modify ideas about the age of the earth and the origins of life if the evidence required such changes, but they constantly used the Bible as one source of respected, true data. For example, when the renowned French naturalist Georges Cuvier (1769–1832) was forced by new fossil discoveries to revise his understanding of the origins of animal and plant life, he proposed that there was not one, but numerous "creations" that could be harmonized with the first chapters of Genesis.[7]

Dana was part of a "school" of nineteenth-century naturalists (later called by speciality names like geologists or zoologists) that was little less than a monument to the task of harmonizing science and the Bible. Known as catastrophists, these disciples of Cuvier believed that multiple natural eruptions or catastrophes, coinciding roughly with the periods of creation in Genesis, were responsible for the character of the earth's geographical and fossil record. Catastrophists were steering a middle course between traditional pre-

---

[5] For Darwin's great admiration of Paley's writings as a student, see Darwin, *Autobiography*, p. 19.
[6] Ibid., pp. 17–26. For the quotations, see the classic study by D. C. Somervell, *English Thought in the Nineteenth Century* (New York: David McKay, 1929), pp. 18–19.
[7] Georges Cuvier, *Essay on the Theory of the Earth* (New York: Kirk and Mercein, 1818), pp. 30–37, 165–174; and John C. Greene, *The Death of Adam* (Ames, Iowa: Iowa State University Press, 1959), ch. 4.

suppositions about the creation and age of the earth (created in six days and being approximately 6000 years old) and a much more radical, nontraditional and less popular position called "uniformitarianism." By assuming that all geological change had occurred by the uniform changes of the earth's surface now in operation, uniformitarians essentially finessed the question of special creative periods and attributed an exceedingly long lifespan to the earth. This position was first argued scientifically by James Hutton (1726–1797), then dramatically confirmed by the work of Charles Lyell (1797–1875) after 1830. Lyell's work was highly controversial, partly because it clashed with the concern of the great majority of scientists before 1830 to harmonize their findings with the Bible. Dana's opinions are similar to other famous naturalists before 1850 like the Englishmen William E. Buckland and the eight writers of the notorious *Bridgewater Treatises* as well as the Americans Benjamin Silliman and Edward Hitchcock.[8]

Second, the reading by Dana illustrates the religious and social roles assumed by most scientists at the time. Before 1859 the great majority of naturalists in England and America were practicing Christians, if not members of the clergy. They regarded their roles as students of "the two books of God"—the special revelation of God in the Bible and the natural revelation of God in nature—as supplementary and interchangeable. Given the constant emphasis on proving the existence of God and demonstrating his characteristics from natural phenomena (called natural theology), thinkers like Dana and William Paley actually, indeed fervently, believed that they were making contributions to theology. In turn, they thought that they were improving and stabilizing the morals of society.

The extent to which natural science was a form of piety and service is captured in the article by Dana. Science was constantly conducted from a theological point of view. It was not yet secularized, that is, considered a nonreligious, worldly enterprise. Nor was it separated from social responsibility. Naturalists frequently popularized their findings from the platform and pulpit alike for the glory of God and the practical good of society.[9]

---

[8] Charles Coultston Gillespie, *Genesis and Geology* (New York: Harper and Brothers, 1951), pp. 201–216; and Conrad Wright, "The Religion of Geology," *New England Quarterly* 14 (June 1941): 335–358.
[9] Cf. Gillispie, *Genesis and Geology,* ch. 7.

Introduction

Charles Darwin's own experience attests to the great significance of the reading from William Paley. Darwin felt that his "careful study" of Paley was one of the few activities at Cambridge that was at all valuable to him educationally. He had been "charmed and convinced" by the classic proof for the existence of God by Paley in *Natural Theology* (1802).[10] In a uniquely articulate way in that book, Paley applied the Newtonian teleological argument—a demonstration that the clock-like order of the solar system proved the existence of a divine Designer—to biological organisms. Paley argued that anatomical organs such as the eye or the hand reflect ingenuity of design and promote happiness far more than do human contrivances such as the telescope or watch, thus proving that these organs were made originally by a benevolent God. This analysis became a first-line defense for theology.

The selection from Paley also reflects the common, pre-Darwinian assumptions concerning the fixity of species and the unchanging character of nature. Paley's argument demonstrated the logical connections between the perfected designs and functions of anatomical organisms and belief in a majestic, divine Creator. He assumed that these creations possessed a degree of perfection appropriate to divine workmanship. Nature and its created species were thus necessarily considered to be developed and complete. Darwin's historicizing of nature as constantly subject to evolution and change was a radical revision of this older conception.

Poetic feeling was the fundamental ingredient separating William Paley from his contemporary William Cowper (1731–1800). Cowper's ever-popular hymn given below plowed Paleyean piety into the religious sensibilities of millions of Christians. In unfathomable mines of anatomical skill the purposes of God were manifest. This religious vision accompanied hundreds of naturalists on their far-reaching excursions, surrounding their investigations with awe. As was true for the youthful Darwin in the midst of a steaming, verdurous forest in Brazil, students of nature were unforgettably filled with "feelings of wonder, admiration, and devotion."[11]

In Cowper's "The Winter Walk at Noon" the ideas concerning teleological proof and nature's perfection found in Paley are further displayed in the context of other themes common at the time. Re-

---
[10] Darwin, *Autobiography*, p. 19.
[11] Ibid., p. 65.

flecting the legacy of numerous Enlightenment figures like Alexander Pope (1688–1744) in England or Jean Jacques Rousseau (1712–1778) in France, Cowper found in nature a realm of order and harmony that offered an ideal standard in contrast to the chaos of human society. Also, in great contrast with the images of nature in the writings of Darwin, William Cowper saw in animal behavior sportive, innocent play—opposed, thought the evangelical and sometimes morosely depressed Cowper, to the cruelty and base gluttony of man.

Through romantic poets like William Wordsworth (1770–1850) and Samuel Taylor Coleridge (1772–1834) in England, and Ralph Waldo Emerson (1803–1884) and Henry David Thoreau (1817–1862) in America, the idealization of nature reached a climax shortly before being severely challenged by the publication of *The Origin of Species.* Beyond Cowper's impressions of animal behavior as sportive and idyllic (even though that behavior for Cowper also reflected the "instinctive fear" and loss of "harmony and family accord" accruing from the sin of Adam and Eve),[12] Wordsworth seemed literally bewildered and shocked by even the cruelty of a robin's chasing a butterfly. Furthermore, though Wordsworth had learned from writers like Paley and Cowper to discover in nature a reflection of God, he went beyond them by finding in the natural world divinity itself. In communion with brooks and hills, his troubled psyche was healed and his soul was infused with a sublime presence. Finally, nature for Wordsworth offered true moral precepts to man. For him who seeks it, love, tenderness and goodness will be so fully learned from nature that,

> abhorrence and contempt are things
> He only knows by name. . . .[13]

A wide range of opinions about the relative goodness or wickedness of human beings was prevalent in the decades before the publication of Darwin's *Origin*. Calvinists from Anglican or Presbyterian or Congregationalist backgrounds believed that although a predes-

---

[12] Brian Spiller, ed., *Cowper* (Cambridge: Harvard University Press, 1968), pp. 524 ff.; and the discussion by Lodwick C. Hartley, *William Cowper* (Chapel Hill: University of North Carolina Press, 1938), ch. 8.
[13] William Wordsworth, *The Complete Poetical Works of William Wordsworth* (Boston: Houghton Mifflin, 1904), p. 462 (from "The Excursion," Bk. IV, lines 1228–1229).

tined minority could reach relative degrees of righteousness through the influence of grace in their lives, inherited depravity made it impossible for humans to perform truly charitable acts. The majority of English and American churchgoers at the time, however, were drifting away from the moorings of their Calvinist ancestors and were opting for Methodist, Baptist, or Low Church Anglican evangelism, or for such forms of Enlightenment religion as orthodox rationalism, or Unitarianism. Against Calvinism, all these groups commonly emphasized that men and women possessed enough intrinsic freedom and goodness to respond personally to the proddings of conscience, the pleadings of evangelists, or the reasoned discourses of liberal clergymen. Influential romantics like Coleridge or Emerson developed these more optimistic beliefs about man still further. Coleridge spoke of an intuitive, divine attribute within the individual, and Emerson revealed softly that man was perfectable to the point of becoming divine.

Quite apart from these religious and metaphysical ideas, prestigious, this-worldly utilitarians like Jeremy Bentham (1748–1832), and James Mill (1773–1836) and his son John Stuart Mill (1806–1873) trusted in mankind enough to believe that if individuals pursued their own interests and pleasures, they would end in producing a reformed and happy society.

This cursory list of attitudes supports the generalization that in spite of the presence of less cheering opinions, there was rising optimism about the goodness of man in England and America before 1859. Marx had not yet made a great impact, Darwin and Freud had not yet spoken, and behind and beyond their ideas, severe social problems had not yet forced the West to recognize its evil and irrational propensities.

On the other hand, aside from all pre-Darwinian assessments of the extent of human virtue and vice, a universal recognition of the immeasurable significance and uniqueness of man prevailed. Several ideas were used to justify this illustrious status of man. Four of these ideas are found in the biblical readings from Psalms and the letter to the Romans, namely, (1) the cardinal presupposition that man was the favored subject of God's concern; (2) the belief that men and women are the special creations of God (Psalm 8 and the earlier reading from Genesis 1–2); (3) that they have control and dominion over all living things (Psalm 8); and (4) that they possess unique mental capacities such as the ability to discover their place in the cosmos. The first of these justifications needs elaboration.

Whether some degree of natural moral goodness was credited to humans (as found in some biblical texts and presupposed by romantics, utilitarians, and most of the religious thinkers above), or whether men and women were considered primarily as self-blinded idolators who are congenitally haughty and ruthless toward others (as in Romans 8), the reigning presupposition in the biblical literature, and consequently in most Western thought, was that the omniscient ruler of the universe was constantly concerned about mankind's welfare. Willingly or unwillingly, man was deeply complimented by religion. And it was everywhere assumed that because religion bestowed intrinsic importance upon man, atheism must mar mankind's image.

The reading from William Kirby (1759–1850) is representative of much English and American scientific and religious opinion about the status of man in the late 1830s. Kirby was one of the eight men chosen to write one of the famous *Bridgewater Treatises* from which the selection in Section I is taken. After five years of work and some 900 pages of discussion concerning the creation, distribution, and anatomical characteristics of biological species, he had, as it were, ascended the organic staircase to man. His belief in the "unhappy *fall*" of Adam and Eve—and hence, all mankind—into sin did not keep him from praising humans as "the beauty of the world," and the "paragon of animals."[14] Notably, Kirby used the concept of the special creation of man with the spirit of God to shore up man's significance in the face of comparative biological data. Man as a rather second-rate biological organism deserves dominion over other species because of his proximity to God established at the time of his origin.

Two additional, popularly utilized justifications for man's unique supremacy that were not listed above in connection with the readings from the Bible appear in Kirby's discussion. These include Kirby's emphasis on the distance of man from the animals—particularly monkeys. This great desire to separate man from monkeys as much as possible, all the while realizing that the newer biological data made it impossible to rule humans out of the animal kingdom

---

[14] The phrases here are taken from a slightly more enthusiastic assessment of man by the English editor of the works of Georges Cuvier, *The Animal Kingdom*, I, ed. by Edward Griffith (London: William Clowes, 1827), p. 152.

Introduction

altogether, was one of the reasons why Kirby and so many other naturalists before Darwin were anxious to embrace the classifications of Cuvier over those of Carolus Linnaeus (1707–1778).[15] Second, and not well developed in Kirby's essay, is an emphasis on mankind's high moral behavior in contrast to animals such as apes. Man's conscience and moral standards present "a striking difference" from other creatures.[16]

Given the evaluations of man represented in the biblical texts and in the selection by William Kirby, one of the most sensitive questions raised by Darwin's work is now clear: Could the old world-view be demolished without debasing mankind?

Section II. Having once believed in the permanence of species and "the strict and literal truth" of the Bible, having accepted at face value William Paley's teleological argument for God, and experienced first hand William Cowper's affectionate reverence for nature, Charles Darwin, through research and writing that was at once exhilarating and laborious, undermined his former beliefs. He substituted evolution by natural selection for the creation and fixity of species, became convinced that the Bible was no longer trustworthy and that miracles were incredible, questioned the principle of design to the point of rejecting it, and claimed that "the grandest scenes" of nature no longer stirred up any of his former feelings of wonder and devotion.[17] Few contrasts are more striking than this catalog of changes. In Darwin's own person an entire world-view was shattered, and a new one, forged out of different materials, was constructed in its place. Darwin of course was not solely responsible for introducing to the West this series of momentous changes, the precise history of which is exceedingly complex. Yet Darwin's work did encourage or demand all of these changes, and Darwin's own intellectual pilgrimage is a crucial, ideal case study indicating why and how these changes occurred in the modern West.

The driving wedge behind these revolutionary revisions in Darwin's thought is succinctly summarized in the phrase "evolution by natural selection." Why did this simply worded concept have such enormous intellectual consequences? How could it, like a demo-

---

[15] Cf. Cuvier, *The Animal Kingdom*, I, pp. 80 ff.; and the discussion in Greene, *Death of Adam*, ch. 6. Versus Linnaeus, Cuvier put men and apes in distinct orders.
[16] Griffith in Cuvier, *The Animal Kingdom*, I, p. 153; and William Whewell, "Astronomy," in *The Bridgewater Treatises*, III (London: William Pickering, 1839), pp. 373 ff.
[17] Darwin, *Autobiography*, pp. 17–18, 65.

lition crane, destroy the structure of the older world-view? Are Darwin's conclusions necessarily the logical ones to draw? These are some of the penetrating questions raised by the readings in this section, all of which are related directly to the topics that have been identified. Certain essential features concerning the meaning and cogency of Darwin's concept of natural selection will highlight the issues discussed in these readings.

The Origin of Species can be described briefly as an elaborate defense of evolution by natural selection as the process that accounts for the origin of life. Throughout the book Darwin constantly opposed the concept of origin by creation with his principle of evolution by natural selection. Of great significance for the intellectual impact of the *Origin* is the fact that it represented the fruition of twenty-eight years of sustained investigation. Darwin's crucial five years of field experience as the official naturalist on board the *Beagle* led into more than twenty years of reading and research unencumbered by financial or domestic stress or the incessant demands of a salaried occupation. His credentials were impressive: vast experience in first-hand observations of geology and nature in its preindustrialized state, disciplined scientific expertise arising out of eight years (and four written volumes) of research on living and fossil barnacles, and twenty years of note-taking and journal-reading in support of the concept of the evolution of species. Not to be underestimated in regard to the *Origin*'s impact was also Darwin's intense personal involvement with his work and his great desire to write clearly and strikingly, a desire admirably achieved (notwithstanding the literary criticism of his work by Karl Marx).[18]

Several crucial ideas that were confirmed or discovered by Darwin were incorporated in the phrase "evolution by natural selection." These ideas help to account for the influence of Darwin's theory. First, in spite of the warnings of his close professorial friend, John Stevens Henslow, Darwin's careful observations of the geological terrain while on his *Beagle* voyage led him to become a "zealous disciple" of Charles Lyell's updated version of uniformitarianism.[19] As opposed to the catastrophic theories of Cuvier and others, Darwin

---

[18] Cf. Charles Darwin, *Life and Letters*, I, ed. by Francis Darwin (New York: D. Appleton, 1897), pp. 513–516; and the brilliant study by Hyman, *The Tangled Bank*, pp. 9–78.
[19] Darwin, *Autobiography*, pp. 36, 143.

Introduction

thus thought of the earth's history in terms of natural, gradual change over an exceedingly long period of time. Given this vast period, small, incremental changes in species could of course amount to great transformations. This nonsupernaturalistic view of the world propounded by Lyell and fully accepted by Darwin was of immeasurable significance for the latter's theoretical perspective.[20]

In spite of the fact that the concept of evolution had been proposed by a number of pre-Darwin writers such as Jean Baptist Lamarck (1744–1829), Darwin's grandfather, Erasmus (1731–1802), and Robert Chambers (1802–1871)—the anonymous writer of the infamous work on evolution, *Vestiges of Creation* (1844)—the credibility of the concept of existing species evolving from preexisting types seemed absurd to the great majority of scientists before 1859.[21] The idea of origin by evolution first struck Charles Darwin because of empirical observations concerning the types of birds, plants, and animals found in different geographical localities. As the *Beagle* sailed to respective continents and the islands close to them, Darwin noted that each separate geographical "province" (a continent and its nearby islands) was inhabited by species with great structural similarities in spite of environmental differences within the given province. For example, species on the Galapagos Islands appeared to be simply varieties of those on the nearest continent, South America. They were not closely related—via some common, created origin—to species in other geographical provinces. Supported by an increasing familiarity with the fossil record, these observations led Darwin to conclude that existing species probably gradually and naturally developed from common ancestors.[22]

The doctrine of special creation merely complicated these empirical observations. Given the facts about the distribution of plant and animal forms, there would have to have been not only multiple creative epochs, but also special creations for each separated province. The logical, natural explanation conceived by Darwin now seemed much more viable than the strained, traditional, supernatural one.

Yet how could evolution occur? Without any understanding of the

---

[20] Ibid., pp. 35, 143–144, 153–155, 164, 176–177; and Huxley in Darwin, *Life and Letters*, I, pp. 543–544.
[21] Cf., e.g., Darwin, *Autobiography*, pp. 178, 183–184, 187–189.
[22] Ibid., pp. 41–42, 150, 186; and Darwin, *Origin of Species* (New York: New American Library, 1958), from the sixth ed. of 1872, chs. 12 and 13.

mechanisms that might lead to the production of new species, evolution merited only the status of an intriguing hunch deserving evaluation. Yet that hunch was attractive enough for Darwin to begin, in July 1837, to collect and record facts bearing on his idea. After immersing himself in "heaps of agricultural and horticultural books" and journals dealing with the breeding and growing of domesticated species, he soon became convinced that "selection" was the "keystone" to the development of new species.[23] But, how could "selection" occur apart from man in nature? The answer came in October of 1838 when Darwin's awareness of the competitive struggle in nature was intensified and given focus by his reading of Thomas Malthus's (1766–1834) *Essay on the Principle of Population.* Malthus demonstrated statistically the great degree to which human populations tend to increase faster than the food supply. At once Darwin recalled that the enormous reproductive capacities of species created a milieu of violent competition for available food. Species born with advantageous variations were thus given a survival edge over their kinsmen and competitors. When prolonged, this process would lead to the formation of new species.[24]

This explanation for the mechanism of evolution was indeed a new, momentous discovery, correctly called "my theory" by Darwin. Species gradually came into being by selective processes that were ordinarily, naturally, inherently present in the natural world. No miraculous creation or providential supervision was needed to account for the origin of species. Building on previous research in geology, biology and philosophy, Darwin secularized the biological world, that is, made it comprehensible apart from religious ideas. Science as he understood it had "nothing to do with Christ."[25] In great contrast to the vast majority of scientists before his time and as a fundamental contribution to later research, Darwin thought that science should shed its theological frame of reference. It could carry on very well without a God hypothesis.

Darwin believed that the necessary correlations to his theory were that the biblical and catastrophic versions of creation were false, that intricate anatomical organisms did not prove the existence

---

[23] Darwin, *Autobiography,* pp. 42, 179, 184; and, appropriately, the first chapter of *The Origin of Species.*
[24] Cf. Darwin, *Autobiography,* pp. 42–43; *The Origin of Species,* pp. 74–77; and *Life and Letters,* I, pp. 437–438, 480.
[25] Darwin, *Autobiography,* p. 61.

of divine activity, and that nature was often "clumsy, wasteful, blundering, low and horribly cruel."[26] The readings from Darwin in this section further define his new opinions and indicate why he believed that "evolution by natural selection" demolished the traditional world-view that was still normative for the great majority of his contemporaries. To what degree was Darwin's own intellectual transformation a necessary corollary for an acceptance of "evolution by natural selection?" And why was Darwin—like Martin Luther or Sigmund Freud—possessed with "a grim willingness to do the dirty work," the lonely but necessary intellectual labor, for his age?[27]

The three sets of readings that follow those by Darwin indicate varying reactions to his ideas and serve both as measures of his cultural impact and as differing critical evaluations of his conclusions. At points, of course, the reader must be alert to the distinction between the direct and the indirect influence of Darwin's own opinions.

The selections from Bishop Samuel Wilberforce (1805–1873) and the then distinguished geologist Adam Sedgwick (1785–1873) illustrate how Darwin's theory of evolution—admittedly gradually, and therefore painlessly appropriated by Darwin[28]—was like a shock wave to many ecclesiastical and scientific traditionalists in 1859. Although uniquely colorful and notable because of their importance and rhetorical abilities, the responses of Wilberforce and Sedgwick actually represented the predominant scientific and religious opinion in the first years after the publication of the *Origin*.[29] At what points were traditional opinions most clearly offended by Darwin's theory of evolution, and what scientific, logical (note the dependence on Paley), and religious grounds were used to attempt to refute Darwin?

Thomas Henry Huxley (1825–1895) and Leslie Stephen (1832–1904) represent articulate and sometimes caustic defenders of Darwin who were in no way inclined to let traditionalists win the day. Huxley contributed greatly to the popularization and institutional support of science, while Stephen publicly proclaimed the honesty and honorableness of free-thinking. Both thinkers sought to assure destiny for science and the demise of religion through their patronage of the celebrated "conflict between science and religion" in the Victorian

---

[26] Darwin as quoted in Greene, *The Death of Adam*, p. 248.
[27] Cf. Erik H. Erikson, *Young Man Luther* (New York: W. W. Norton, 1962), pp. 8–9.
[28] Darwin, *Autobiography*, p. 62.
[29] Cf. Huxley in Darwin, *Life and Letters*, I, p. 540.

era. As graphically portrayed in the *Puck* drawing included on page 104, science was pictured as the triumphant banisher of antiquated, superstitious, fear-instilling beliefs of the past—equated particularly with religion. Science was ushering in a new era of progress because of its discoveries of new truth about man, nature, and society, and because, beneath these discoveries, it possessed the only reliable method for obtaining truth. Science was thus identified with progress and change, which were increasingly regarded in the nineteenth century as fundamental to the very nature of history. Religion was given an opposite image. At best it passively and reluctantly accepted inevitable change. It could never become an innovative social or intellectual force. At worst, it opposed truth with all the bigotry and influence that it could muster.[30]

The readings from Asa Gray (1810–1888), Baden Powell (1796–1860) and S. R. Driver (1846–1914) illustrate responses to Darwin and modern science quite different from reactionaries like Wilberforce and Sedgwick or supporters like Huxley and Stephen. Gray, Powell, and Driver accepted Darwin's biological research as true, but questioned the adequacy of religious and philosophical conclusions like his. Gray sympathetically advocated Darwin's biological theory in America, but like a number of other writers of the time—Charles Kingsley (1819–1875) and Henry Drummond (1851–1897) in England, or John Fisk (1842–1901) and Lyman Abbot (1835–1922) in America—believed that Darwinism did not necessitate a rejection of teleology, even if it did force a reconsideration of the particular arguments of William Paley. The reader can decide about the validity of Gray's contentions.

The selections from Baden Powell and S. R. Driver lead to a critical evaluation of the assumption that all religion should be equated with traditionalism and should necessarily expire at the sound of the voice of science. Darwinism was one among other developments—scientific, philosophical and historical—influencing Powell and Driver. Like Immanual Kant (1724–1804), Frederick Schleiermacher (1768–1834), Samuel Taylor Coleridge, and Ralph Waldo Emerson, Powell proposed that religion did not and should not depend on

---

[30] Cf., e.g., the readings by John Dewey and Andrew Dickson White in Philip Appleman, *Darwin* (New York: W. W. Norton, 1970), pp. 393–402, 432–438. For the ecstatic faith in science, see Walter E. Houghton, *The Victorian Frame of Mind* (New Haven: Yale University Press, 1957), pp. 33–45.

logical, historical, or scientific proofs for the existence of God. The article by Driver illustrates a typical response by liberal Jews and Christians to the assumption that Darwinian science had discredited the Bible. As one of a growing number of biblical critics, Driver demonstrated that the biblical literature was written against the background of ancient Near Eastern thought and manifested no original intention of being scientifically infallible. He concluded that the Bible thus did not lose its significance when proved factually or historically inaccurate. Is his demonstration convincing and the conclusion warranted?

The sensitive, intriguing poems of Alfred Lord Tennyson (1809–1892) and Thomas Hardy (1840–1928) compose a dramatic critique of the beliefs and feelings of Paley, Cowper, Wordsworth, and the generations of poets who had found in nature sure signs of peace, harmony, and faith. The apparent silence of nature belies the shrieking horror of innumerable years of suffering, violence, and death. The apparent calm of nature is but an illusion for the constant, brutal battle for food and livelihood. Rather than offering moral precepts, nature seemed indifferent to any conduct other than instinctive survival for the physically fit, not the morally best.

In contrast to Cowper and Wordsworth, Hardy could not find in the woods a restful release from fretful humanity. Only by reversing this movement, only by turning from nature to humankind, was Hardy able to find comfort and healing. In further contrast with Paley, Cowper, and Wordsworth, neither Hardy nor Tennyson discovered in nature signs of divine purpose. Hardy could not discover in nature any of the former hope invoked by the song of a thrush at evening. Tennyson's reactions were complex, reflecting in part the confusions and doubts of the Victorian era. He still believed in God (although questioning whether God might be "They" or "One" or "All," rather than the orthodox "He") but thought that nature's continual, purposeless strife created a grave problem for faith in a loving deity. So Tennyson sought God in the immortality of the soul and in human moral consciousness, by the standards of which he charged nature with the crimes of disease, murder and appalling waste. Nevertheless, although forsaking Paley and his disciples, Tennyson saw in nature incredible beauty, impressive order and a kind of inexplicable vitalism. Nature had fallen from grace, but had not yet surrendered all its mysteries to science.

Section III. Not long after Charles Darwin first sensed that life originated by evolution instead of creation, he wrote that man had arisen out of an animal past and was not the favorite creation of a loving God. In his early notebook of 1837 on transmutation he spoke of animals as "our fellow brethren in pain, disease, death, suffering, and famine" with which we are "all melted together" because we originate from "one common ancestor."[31] Shortly after the publication of *The Origin of Species* twenty-two years later, Darwin wrote to Charles Lyell, affirming that it was "impossible to doubt" that man was in "the same predicament as other animals." Man has the "pleasant genealogy" of a fish-like hermaphrodite for an ancestor.[32]

Darwin's mature study of man appeared with the publication of *The Descent of Man* in 1871, from which the reading in this section is taken. In *Descent* Darwin sought to account for man's total development—physical, mental, moral, social, and religious—as wholly dependent upon forces naturally and ordinarily operative in nature. In this he was both elaborating logically the point of view manifested in *The Origin of Species* and reflecting in a thoroughgoing way the naturalistic understanding of man and nature that had been developing for some time.

Darwin believed that human development was primarily indebted to "natural selection," that is, the competitive struggle for survival in nature, but he also expanded the significance of other mechanisms, chiefly "sexual selection." Darwin argued that even as the choosing of mates by many animal species led to the acquisition of such characteristics as bright coloration, complex courting behavior, and horns and teeth as weapons, so also mate-selection by humans led to the development of physical features like relatively hairless skin and, for females, well-developed breasts. Darwin also thought that races evolved primarily because of sexual selection. Nevertheless, although he did not believe that the competition of natural selection wholly accounted for man, Darwin steadily maintained that all the mechanisms responsible for human evolution were processes endemic to the natural world. Man in no way transcended the forces of the earth from which he was formed. No mystical or supernatural influences were entertained.

In order to make the strongest possible case for the natural and

---

[31] Darwin, *Autobiography*, p. 179.
[32] Darwin, *Life and Letters*, II, pp. 59–60.

Introduction

animal origins of man in *Descent,* Darwin sought to obliterate systematically the vivid, quantitative distinctions previously drawn between humans and animals. To the core of his being, man is an animal. An analysis of his embryonic development, vestigial organs, bodily structure, and emotions point conclusively to his animal pedigree, Darwin argued. Man differs from his animal brethren only in degree, not in kind. Darwin thus opposed Georges Cuvier and the vast majority of biologists in the first half of the nineteenth century who had endorsed Cuvier's classification of man into a biological order of his own. Man deserves the rank of "merely a Family, or possibly even only a Sub-family" within the order of the primates.[33]

In spite of the fact that he obliterated many human-animal distinctions and considered man's development to be dependent on solely natural mechanisms, Darwin spoke highly, at times quite ecstatically, about mankind. But having undermined three of the traditional supports for human dignity, evidenced prominently in the readings in Section I—man's position as the center of divine or providential concern, his unique creation "in the image of God," and his enormous distance from animals and apes—how could Darwin justify regarding humans as "the wonder and glory of the Universe" who should not "feel ashamed" of their animal origins?[34] This is one of the critical questions put to the reader of the selection from *The Descent of Man,* and it is a question intensely debated by Darwin's critics and defenders.

The readings from Bishop Wilberforce and Adam Sedgwick in Section II indicate that religious and biological traditionalists resolutely believed that Darwinism debased man. They regarded traditional religion as an indispensable factor in preserving human dignity. To sever humanity's ties with God was to "brutalize" and "sink" the human race into its lowest state of "degradation," cried Sedgwick.[35] This belief represents a constant theme among traditionalists from the time of Sedgwick in 1859 to William Jennings Bryan at the Scopes trial in 1925 to the present time. Darwinism was regarded as tantamount to atheism, and atheism debases man. Is there validity in either or both of these charges?

Apart from traditionalists, differing evaluations about Darwin's

---

[33] Darwin, *Descent of Man,* pp. 123–155.
[34] Ibid., pp. 168–169.
[35] Darwin, *Autobiography,* p. 229.

opinions concerning man and his origins appeared among avowed evolutionists. The readings from Alfred Russel Wallace (1823–1913) and John Fiske (1842–1901) represent two of these evaluations.

Wallace's renown was based on his reputation as a "co-discoverer" with Darwin of the principle of evolution by natural selection, as well as on his work in shaping the modern study of the geographical distribution of animals. Yet in spite of his commitment to the great significance of evolution by natural selection and of his scientific expertise, Wallace could not endure Darwin's conclusion that natural mechanisms alone could account for mankind's intellectual abilities and aesthetic and moral faculties. Did not the "immense" differences between the *lowest man* and the *highest ape,* which even Darwin admitted, suggest that a new power or "vitality," which was not gradually and naturally abstracted from the lower animals, had been operative in the development of man?[36] Decrying the "hopeless and soul-deadening belief" inherent in Darwin's conviction that man's evolution was wholly natural, Wallace claimed that the mysteries of life evidenced in man's complex behavior and beliefs pointed to the influence of some transcendent "spiritual influx."[37] In both his negative reactions to Darwin and in his arguments favoring the influence of some mysterious, nonmaterial power in the evolution of man, Wallace represented the reactions of other evolutionists, notably the Catholic biologist St. George Mivart (1827–1900) and the man who had influenced Darwin profoundly, Charles Lyell.[38]

John Fiske began with the revisions of Darwin by Wallace and Mivart and, along with a number of other thinkers such as Henry Drummond in England, and Minot Judson Savage (1841–1918) and Lyman Abbot (1835–1922) in America, depicted the entire history of evolution in a remarkably new light. Instead of focusing, as Darwin had, on man's proximity to the animals—thus causing man to "descend" from his former divinely created status into the lap of nature—Fiske took as his interpretative key the idea that evolution is the history of the *ascent* of sensate life from amoeba to man. By focusing on developments in cephalization (evolutionary specialization through the nervous system and brain), Fiske argued that man, possessing the

---

[36] Alfred Russel Wallace, *Darwinism* (London: Macmillan, 1891), pp. 474 ff.
[37] Ibid., pp. 476–477.
[38] Cf. St. George Mivart, *On the Genesis of Species* (New York: D. Appleton, 1871); and Charles Lyell, *The Geological Evidences of the Antiquity of Man* (Philadelphia: George W. Childs, 1863).

most complex neural system on earth, was a "masterpiece" of the evolutionary process. Furthermore, he believed that the process of cephalization leading up to man displayed a directional force that had moral and cosmic dimensions. "Spiritual perfection" is the goal to which all evolution points, and that goal is reached not by some kind of supernatural infusion of Spirit (as is more the case for Wallace and Mivart) but by the gradual operation of a divine force immanent in nature itself. Thus for Fiske, evolution served both to glorify man as the "wonder and glory of the Universe" and to instill hope in the "triumphant" ethical future.[39] Fiske's opinions are remarkably similar to the evolutionary mysticism of Teilhard de Chardin (1881–1955).[40]

In T. H. Huxley, Darwinism secured the services of an unconquerable controversialist. With impressive intellectual agility, Huxley time and again identified points where Darwinism conflicted with traditional and quasi-traditional opinions, then sought to vindicate Darwinism against the disparaging aspersions of its critics. Huxley did not mince words, nor did he cloud the issues with obscure or aimless rhetoric. He thought that the implications of Darwin's research should be frankly stated, openly discussed, and of course for Huxley, readily accepted. His writings thus become excellent channels through which the cultural impact and personal meaning of Darwinism can be assessed.

The readings from Huxley answer and pose questions. First, are the opinions of Alfred Russel Wallace friendly to the logical implications of Darwinism? Huxley's negative answer illustrates his and Charles Darwin's thoroughgoing naturalism, and poses questions as to whether Wallace or Huxley got the better of the argument or manifested the most legitimate perspective. Given the directional force of evolution identified by Fiske, can a degree of cosmic purpose be discovered in the achievements of natural selection? Huxley did not

---

[39] Both of the quoted phrases are from Darwin's *Descent of Man* and indicate how Fiske drew upon aspects of Darwin's own thought. Cf. Darwin, *Descent,* pp. 127, 167–168; and John Fiske, *Through Nature to God* (Boston: Houghton Mifflin, 1899), pp. 113 ff.

[40] Cf. Teilhard de Chardin, *The Phenomenon of Man* (New York: Harper and Row, 1961). For a discussion comparing Fiske and other nineteenth-century thinkers with Teilhard de Chardin, see Ernst Benz, *Evolution and Christian Hope* (Garden City, New York: Doubleday, 1966), ch. 9. For Pierre de Chardin with bibliography, cf. H. Paul Santmire in *Critical Issues in Modern Religion,* Roger A. Johnson et al. (Englewood Cliffs, N. J.: Prentice-Hall, 1973), ch. 4.

say, but the dedication of John Fiske's book (from which the reading was taken) to "the beloved and revered memory of my friend Thomas Henry Huxley" implies that Fiske's ideas probably received a warmer reception from Huxley than the opinions of Wallace.

Second, does the secular development of man—outlined by Darwin—as endorsed and expanded by Huxley, lead to the degradation of man? No, answered Huxley, challenging the reader to evaluate his reasons in the light of previous criticisms of Darwin. Third, given Huxley's and Darwin's revolutionary beliefs that social institutions, morals and religion are natural, human developments, what was to be done with traditional thought and literature that presupposed mystical and supernatural origins for these developments? Huxley's conclusion that the Bible itself should continue to serve as a source of insight and social reform is in keeping with his overall conviction that an agnostic was a friend, not a foe of man's cultural past.

The poems from Thomas Hardy and Algernon Charles Swinburne (1837–1909) represent the polarities of loneliness and liberation characteristically manifested by those who read Darwin.[41] Hardy, like Alfred Russel Wallace, or the poets Matthew Arnold (1822–1888), Arthur Hugh Clough (1819–1861), and James Thomson (1834–1882) found in evolutionary naturalism melancholy despair. The loss of security and faith exposed and isolated humanity, causing man, as poignantly depicted in the picture by Edvard Munch (p. 193), to cry out in the midst of his earthly hell.

Swinburne was closer to Huxley and represents poets like Edward Fitzgerald (1809–1883) and George Meredith (1828–1909), for whom Darwinism was a source of liberation from the tyrannies of the past. Arising out of the green earth, mankind, aided by science, was emancipated from a parasitic God and freed for a confident future.

For many others, Alfred Lord Tennyson was the true voice of mankind's emotional response to a post-Darwinian status. As the reading from Section II indicates, Tennyson steered between the polarities of loneliness and liberation with the aid of a trust that fused doubt, expectation and cosmic purpose.

The final readings from Herbert Spencer (1820–1903) and T. H. Huxley represent classic social responses to evolutionary thought.

---

[41] These are the emotional attitudes perceptively highlighted by Walter E. Houghton, *The Victorian Frame of Mind,* chs. 2 and 3, and previously identified by Lionel Stevenson in *Darwin Among the Poets* (Chicago: University of Chicago Press, 1932).

Introduction

For Spencer in England and his numerous American disciples—notably William Graham Sumner (1840–1910)—the concept of evolution through the struggle for survival became the key that unlocked the secrets behind human social development. Thanks to evolution by natural selection, human society slowly and steadily progressed from hostile, regimented, primitive communities to peaceful, interdependent, complex ones. And the way to assure that further progress occurs is to make sure that the stern, self-operating, natural "laws" of individualism and competition continue to eliminate the unfit and to place future power and influence in the hands of the most intelligent and talented. Laissez-faire social and political policies were thus equated with Darwinian laws of nature and regarded by Spencer and his disciples as leading inexorably and naturally to a better society. Social reform schemes were naturally regarded as "artificial" interferences leading to social retrogression.

Spencerism became enormously popular and influential in America and served as a significant political option in England. It was accepted as portraying in a perfectly accurate way the exploitative conditions of competitive industrial societies. Furthermore, Spencer's thought served as an historical, natural, if not cosmic, legitimation of laissez-faire society. Standing as they were at the top of the American financial and social ladders, men such as Andrew Carnegie and John D. Rockefeller heartily approved of Spencer's political philosophy. There is hardly a clearer example of self-justifying adulation.

The last reading, by T. H. Huxley, shows that there was no single "Social Darwinism," that is, no single application of the principles of Darwinian evolution to society. The Russian, Peter Kropotkin (1842–1921) and the Englishman Henry Drummond utilized Darwin's research to show that mutual aid and "altruism," not just competition, were of fundamental importance in the natural world. Huxley, like the American social reformer Lester Ward (1841–1910), rejected resignation to the inexorable laws of nature as recommended in the writings of laissez-faire Darwinists like Spencer. Man also is a product of evolution, argued Huxley and Ward. And humans must use their intelligence and hard-won, slowly developed ethical standards to modify the conditions of their existence. To be fully human is "to strive, to seek, to find, and not to yield."

The Clash of Issues

The origin of life:
> In the beginning God created the Heavens and the Earth.
> GENESIS 1:1

> The view which most naturalists until recently entertained, and which I formerly entertained—namely, that each species has been independently created—is erroneous.
> CHARLES DARWIN

Proof of God from design in nature:
> The marks of *design* [in nature] are too strong to be gotten over. Design must have had a designer. That designer must have been a person. That person is God.
> WILLIAM PALEY

> The old argument from design in Nature, as given by Paley . . . fails, now that the law of natural selection has been discovered.
> CHARLES DARWIN

> The adoption . . . of Darwin's particular hypothesis . . . would leave the doctrine of . . . design, just where [it was] before.
> ASA GRAY

Nature:
> Therefore am I still . . . well pleased to recognize
> In nature . . .
> The anchor of my purest thoughts, the nurse,
> The guide, the guardian of my heart. . . .
> WILLIAM WORDSWORTH

> What a book a devil's chaplain might write on the clumsy, wasteful, blundering, low, and horribly cruel works of nature!
> CHARLES DARWIN

Man:
> Man's derived supremacy over the earth . . . man's gift of reason . . . are . . . utterly irreconcilable with the degrading notion of the brute origin of him who was created in the image of God. . . .
> SAMUEL WILBERFORCE

> Thoughtful men . . . will find in the lowly stock whence Man has sprung, the best evidence of the splendor of his capacities. . . .
> T. H. HUXLEY

Society:
> Society . . . cannot without . . . disaster interfere with the play of . . . principles under which every species has reached such fitness . . . as it possesses. . . .
> HERBERT SPENCER

> The ethical progress of society depends, not on imitating the cosmic process . . . but in combating it.
> T. H. HUXLEY

I BEFORE THE ORIGIN OF SPECIES (1859)

Genesis 1:1–2:9, 2:15–25
THE BIBLICAL STORY OF CREATION

The Bible begins with this narrative of creation, which tells how the heavens, earth, species and man were miraculously and directly formed by God in the course of a seven-day Hebrew week. Each part of nature is called "good," except for man, who is seen as "very good," and the Hebrews, followed by the West, presupposed that nature "declared the glory of God." Until the beginning of scientific geology in the eighteenth century, there was little reason on scientific grounds to question the literal sense of the story. In fact, based on the work of the Irish archbishop James Ussher (1581–1656), many English Bibles put in their margins next to these chapters the date 4004 B.C. What is the image of man in the story? Does there appear to be one account of the creation or two (Genesis 1:1–2:4a, 2:4b ff.)? Might the intrinsic character of the narratives as well as the observation that two, differing stories were placed back-to-back suggest that strict, historical accuracy was absent from the Hebraic world-view?

1

In the beginning God created[a] the heavens and the earth. ²The earth was without form and void, and darkness was upon the face of the deep; and the Spirit[b] of God was moving over the face of the waters.

3 And God said, "Let there be light"; and there was light. ⁴And God saw that the light was good; and God separated the light from the darkness. ⁵God called the light Day, and the darkness he called Night. And there was evening and there was morning, one day.

6 And God said, "Let there be a firmament in the midst of the waters, and let it separate the waters from the waters." ⁷And God made the firmament and separated the waters which were under the firmament from the waters which were above the firmament. And it was so. ⁸And God called the firmament Heaven. And there was evening and there was morning, a second day.

9 And God said, "Let the waters under the heavens be gathered together into one place, and let the dry land appear." And it was so. ¹⁰God called the dry land Earth, and the waters that were gathered to-

From the Revised Standard Version Bible, copyright 1946, 1952 and © 1971, and used by permission.

[a] Or *When God began to create.*
[b] Or *wind.*

gether he called Seas. And God saw that it was good. ¹¹And God said, "Let the earth put forth vegetation, plants yielding seed, and fruit trees bearing fruit in which is their seed, each according to its kind, upon the earth." And it was so. ¹²The earth brought forth vegetation, plants yielding seed according to their own kinds, and trees bearing fruit in which is their seed, each according to its kind. And God saw that it was good. ¹³And there was evening and there was morning, a third day.

14 And God said, "Let there be lights in the firmament of the heavens to separate the day from the night; and let them be for signs and for seasons and for days and years, ¹⁵and let them be lights in the firmament of the heavens to give light upon the earth." And it was so. ¹⁶And God made the two great lights, the greater light to rule the day, and the lesser light to rule the night; he made the stars also. ¹⁷And God set them in the firmament of the heavens to give light upon the earth, ¹⁸to rule over the day and over the night, and to separate the light from the darkness. And God saw that it was good. ¹⁹And there was evening and there was morning, a fourth day.

20 And God said, "Let the waters bring forth swarms of living creatures, and let birds fly above the earth across the firmament of the heavens." ²¹So God created the great sea monsters and every living creature that moves, with which the waters swarm, according to their kinds, and every winged bird according to its kind. And God saw that it was good. ²²And God blessed them, saying, "Be fruitful and multiply and fill the waters in the seas, and let birds multiply on the earth." ²³And there was evening and there was morning, a fifth day.

24 And God said, "Let the earth bring forth living creatures according to their kinds: cattle and creeping things and beasts of the earth according to their kinds." And it was so. ²⁵And God made the beasts of the earth according to their kinds and the cattle according to their kinds, and everything that creeps upon the ground according to its kind. And God saw that it was good.

26 Then God said, "Let us make man in our image, after our likeness; and let them have dominion over the fish of the sea, and over the birds of the air, and over the cattle, and over all the earth, and over every creeping thing that creeps upon the earth." ²⁷So God created man in his own image, in the image of God he created him; male and female he created them. ²⁸And God blessed them, and God

The Biblical Story of Creation

said to them, "Be fruitful and multiply, and fill the earth and subdue it; and have dominion over the fish of the sea and over the birds of the air and over every living things that moves upon the earth." [29]And God said, "Behold, I have given you every plant yielding seed which is upon the face of all the earth, and every tree with seed in its fruit; you shall have them for food. [30]And to every beast of the earth, and to every bird of the air, and to everything that creeps on the earth, everything that has the breath of life, I have given every green plant for food." And it was so. [31]And God saw everything that he had made, and behold, it was very good. And there was evening and there was morning, a sixth day.

2

Thus the heavens and the earth were finished, and all the host of them. [2]And on the seventh day God finished his work which he had done, and he rested on the seventh day from all his work which he had done. [3]So God blessed the seventh day and hallowed it, because on it God rested from all his work which he had done in creation.

4 These are the generations of the heavens and the earth when they were created.

In the day that the *Lord* God made the earth and the heavens, [5]when no plant of the field was yet in the earth and no herb of the field had yet sprung up—for the *Lord* God had not caused it to rain upon the earth, and there was no man to till the ground; [6]but a mist[c] went up from the earth and watered the whole face of the ground— [7]then the *Lord* God formed man of dust from the ground, and breathed into his nostrils the breath of life; and man became a living being. [8]And the *Lord* God planted a garden in Eden, in the east; and there he put the man whom he had formed. [9]And out of the ground the *Lord* God made to grow every tree that is pleasant to the sight and good for food, the tree of life also in the midst of the garden, and the tree of the knowledge of good and evil. . . .

15 The *Lord* God took the man and put him in the garden of Eden to till it and keep it. [16]And the *Lord* God commanded the man, saying, "You may freely eat of every tree of the garden; [17]but of the tree of the knowledge of good and evil you shall not eat, for in the day that you eat of it you shall die."

[c] Or *flood.*

18 Then the *Lord* God said, "It is not good that the man should be alone; I will make him a helper fit for him." ¹⁹So out of the ground the *Lord* God formed every beast of the field and every bird of the air, and brought them to the man to see what he would call them; and whatever the man called every living creature, that was its name. ²⁰The man gave names to all cattle, and to the birds of the air, and to every beast of the field; but for the man there was not found a helper fit for him. ²¹So the *Lord* God caused a deep sleep to fall upon the man, and while he slept took one of his ribs and closed up its place with flesh; ²²and the rib which the *Lord* God had taken from the man he made into a woman and brought her to the man. ²³Then the man said,

> "This at last is bone of my bones
> and flesh of my flesh;
> she shall be called Woman,
> because she was taken out of Man."

²⁴Therefore a man leaves his father and his mother and cleaves to his wife, and they become one flesh. ²⁵And the man and his wife were both naked, and were not ashamed.

J. J. S. Perowne
THE CHARACTER OF RELIGION

When J. J. S. Perowne (1823–1904) was chosen to write several articles concerning the Old Testament in William Smith's A Dictionary of the Bible, *he could be certain that he would have a wide, interdenominational reading audience for forty years. Smith's four-volumed* Dictionary *passed through several editions and reprintings and became the standard reference work in the English language from 1860 to 1900. As an Anglican known for his scholarship in biblical literature and for his position as vice-principal of St. David's College in Wales, Perowne joined other Anglicans (Protestant Episcopalians in America), Congregationalists, Presbyterians, and Baptists in composing a dictionary that proposed to combine scholarship and faith.*

His article on the Book of Genesis, from which the reading here is taken, reflects the purpose of defending the integrity of the Bible in light of pre-Darwinian science. Perowne accents the historical trustworthiness of Genesis and claims that since the "days" of creation are to be interpreted as creative eras, there is no conflict between Genesis and geology. Nevertheless, he says that the religious purpose and "popular language" of the Bible, although "in no way opposed" with science, mean that only general agreement between the two is to be expected.

The book of Genesis has an interest and an importance to which no other document of antiquity can pretend. If not absolutely the oldest book in the world, it is the oldest which lays any claim to being a trustworthy history. There may be some papyrus-rolls in our museums which were written in Egypt about the same time that the genealogies of the Semitic race were so carefully collected in the tents of the Patriarchs. But these rolls at best contain barren registers of little service to the historian. It is said that there are fragments of Chinese literature which in their present form date back as far as 2200 years B.C., and even more. But they are either calendars containing astronomical calculations, or records of merely local and temporary interest. Genesis, on the contrary, is rich in details respecting other races besides the race to which it more immediately belongs. And the Jewish pedigrees there so studiously preserved are but the scaffolding whereon is reared a temple of universal history.

If the religious books of other nations make any pretensions to

Abridged from J. J. S. Perowne, "Genesis," in *A Dictionary of the Bible,* I, ed. by William Smith (Boston: Little, Brown, 1860).

vie with it in antiquity, in all other respects they are immeasurably inferior. The Mantras, the oldest portions of the Vedas, are, it would seem, as old as the fourteenth century B.C. The Zendavesta, in the opinion of competent scholars, is of very much more modern date. Of the Chinese sacred books, the oldest, the Yih-king, is undoubtedly of a venerable antiquity, but it is not certain that it was a religious book at all; while the writings attributed to Confucius are certainly not earlier than the sixth century B.C.

But Genesis is neither like the Vedas, a collection of hymns more or less sublime; nor like the Zendavesta, a philosophic speculation on the origin of all things; nor like the Yih-king, an unintelligible jumble whose expositors could twist it from a cosmological essay into a standard treatise on ethical philosophy. It is a history, and it is a religious history. The earlier portion of the book, so far as the end of the eleventh chapter, may be properly termed a history of the world; the latter is a history of the fathers of the Jewish race. But from first to last it is a religious history: it begins with the creation of the world and of man; it tells of the early happiness of a Paradise in which God spake with man; of the first sin and its consequences; of the promise of Redemption; of the gigantic growth of sin, and the judgment of the Flood; of a new earth, and a new covenant with man, its unchangeableness typified by the bow in the heavens; of the dispersion of the human race over the world. And then it passes to the story of Redemption; to the promise given to Abraham, and renewed to Isaac and to Jacob, and to all that chain of circumstances which paved the way for the great symbolic act of Redemption, when with a mighty hand and a stretched-out arm Jehovah brought his people out of Egypt.

Hard critics [of Genesis] have tried all they can to mar its beauty and to detract from its utility. In fact the bitterness of the attacks on a document so venerable, so full of undying interest, hallowed by the love of many generations, makes one almost suspect that a secret malevolence must have been the mainspring of hostile criticism. Certain it is that no book has met with more determined and unsparing assailants. To enumerate and to reply to all objections would be impossible. We will only refer to some of the most important.

The story of Creation, as given in the first chapter, has been set aside in two ways: first by placing it on the same level with other cosmogonies which are to be found in the sacred writings of all

The Character of Religion

nations; and next, by asserting that its statements are directly contradicted by the discoveries of modern science.

Let us glance at these two objections.

Now when we compare the Biblical with all other known cosmogonies, we are immediately struck with the great *moral* superiority of the former. There is no confusion here between the Divine Creator and His work. God is before all things, God creates all things; this is the sublime assertion of the Hebrew writer. Whereas all the cosmogonies of the heathen world err in one of two directions. Either they are Dualistic, that is, they regard God and matter as two eternal co-existent principles; or they are Pantheistic, i.e. they confound God and matter, making the material universe a kind of emanation from the great Spirit which informs the mass. Both these theories, with their various modifications, whether in the more subtle philosophies of the Indian races, or in the rougher and grosser systems of the Phoenicians and Babylonians, are alike exclusive of the idea of creation. Without attempting to discuss in anything like detail the points of resemblance and difference between the Biblical record of creation, and the myths and legends of other nations, it may suffice to mention certain particulars in which the superiority of the Hebrew account can hardly be called in question. First, the Hebrew story alone clearly acknowledges the personality and unity of God. Secondly, here only do we find recognized a distinct act of creation, by creation being understood the calling into existence out of nothing the whole material universe. Thirdly, there is here only a clear intimation of that great law of progress which we find everywhere observed. The *order* of creation as given in Genesis is the gradual progress of all things from the lowest and least perfect to the highest and most completely developed forms. Fourthly, there is the fact of a relation between the personal Creator and the work of His fingers, and that relation is a relation of Love: for God looks upon His creation at every stage of its progress and pronounces it very good. Fifthly, there is throughout a sublime simplicity, which of itself is characteristic of a history, not of a myth or of a philosophical speculation.

It would occupy too large a space to discuss at any length the objections which have been urged from the results of modern discovery against the literal truth of this chapter. One or two remarks of a general kind must suffice. It is argued, for instance, that light could not have existed before the sun, or at any rate not that kind of light which would be necessary for the support of vegetable life; whereas the Mosaic narrative makes light created on the first day, trees and plants on the third, and the sun on the fourth. To this we may reply, that we must not too hastily build an argument upon our ignorance. We do not *know* that the existing laws of creation were in operation when the creative fiat was first put forth. The very act of Creation must have been the introducing of laws: but when the work was finished, those laws may have suffered some modification. Men are not now created in the full stature of manhood, but are born and grow. Similarly the lower ranks of being might have been influenced by certain necessary conditions during the first stages of their existence, which conditions were afterwards removed without any disturbance of the natural functions. And again it is not certain that the language of Genesis can only mean that the sun was *created* on the fourth day. It *may* mean that then only did that luminary become visible to our planet.

With regard to the six days, no reasonable doubt can exist that they ought to be interpreted as six periods, without defining what the length of those periods is. No one can suppose that the Divine rest was literally a rest of twenty-four hours. On the contrary, the Divine Sabbath still continues. There has been no *creation* since the creation of man. This is what Genesis teaches, and this geology confirms. But God, after six periods of creative activity, entered into that Sabbath in which His work has been not a work of Creation but of Redemption.

No attempt, however, which has as yet been made to identify these six periods with corresponding geological epochs can be pronounced satisfactory. On the other hand, it seems rash and premature to assert that no reconciliation is possible. What we ought to maintain is, that no reconciliation is necessary. It is certain that the author of the first chapter of Genesis, whether Moses or someone else, knew nothing of geology or astronomy. It is certain that he made use of phraseology concerning physical facts in accordance with the limited range of information which he possessed. It is also certain that the Bible was never intended to reveal to us knowledge of which our own faculties rightly used could put us in possession. And we have no business therefore to expect anything but popular language in the description of physical phenomena. Thus, for instance, when it is

The Character of Religion

said that by means of the firmament God divided the waters which were above from those which were beneath, we admit the fact without admitting the implied explanation. The *Hebrew* supposed that there existed vast reservoirs above him corresponding to the "waters under the earth." *We* know that by certain natural processes the rain descends from the clouds. But the *fact* remains the same that there are waters above as well as below.

Further investigation may perhaps throw more light on these interesting questions. Meanwhile it may be safely said that modern discoveries are in no way opposed to the great outlines of the Mosaic cosmogony. That the world was created in six periods, that creation was by a law of gradual advance beginning with inorganic matter, and then advancing from the lowest organisms to the highest, that since the appearance of man upon the earth no new species have come into being; these are statements not only not disproved, but the two last of them at least amply confirmed by geological research.

Another fact which rests on the authority of the earlier chapters of Genesis, the derivation of the whole human race from a single pair, has been abundantly confirmed by recent investigations. For the full proof of this it is sufficient to refer to Prichard's *Physical History of Mankind,* in which the subject is discussed with great care

James D. Dana
THE CHARACTER OF SCIENCE

Given the overt religious intent of this reading, it may seem astonishing that it is chosen from the last chapter of a popular textbook in geology by the famous American geologist and zoologist, James D. Dana (1813–1895). Dana was the professor of natural history at Yale from 1849 to 1890, and in 1862 he published the first edition of his Manual of Geology, *from which this excerpt entitled "Cosmogony" is taken. Like numerous other scientists in England and America at the time, Dana believed not only that science and the Bible could be fully harmonized, but also that his discipline confirmed the accuracy and divine character of scripture. Not until 1883 did Dana announce publicly his acceptance of a greatly modified, religiously informed version of Darwin's theory of evolution.*

The science of cosmogony treats of the history of creation.

Geology comprises that later portion of the history which is within the range of direct investigation, beginning with the rock-covered globe, and gathering only a few hints as to a previous state of igneous fluidity.

Through Astronomy our knowledge of this earlier state becomes less doubtful, and we even discover evidence of a period still more remote. Ascertaining thence that the sun of our system is in intense ignition, that the moon, the earth's satellite, was once a globe of fire, but is now cooled and covered with extinct craters, and that space is filled with burning suns—and learning also from physical science that all heated bodies in space must have been losing heat through past time, the smallest most rapidly—we safely conclude that the earth has passed through a stage of igneous fluidity.

Again, as to the remoter period: the forms of the nebulae and of other starry systems in the heavens, and the relations which subsist between the spheres in our own system, have been found to be such as would have resulted if the whole universe had been evolved from an original nebula or gaseous fluid. . . . If, then, this nebular theory be true, the universe has been developed from a primal unit, and the earth is one of the individual orbs produced in the course of its evolution. Its history is in kind like that which has been deciphered with

From James D. Dana, *Manual of Geology* (Philadelphia: Theodore Bliss, 1865).

The Character of Science

regard to the earth: it only carries the action of physical forces, under a sustaining and directing hand, further back in time.

The science also of Chemistry is aiding in the study of the earth's earliest development, and is preparing itself to write a history of the various changes which should have taken place among the elements from the first commencement of combination to the formation of the solid crust of our globe.

It is not proposed to enter either into chemical or astronomical details in this place, but, supposing the nebular theory to be true, briefly to mention the great stages of progress in the history of the earth, or those successive periods which stand out prominently in time through the exhibition of some new idea in the grand system of progress. The views here offered, and the following on the cosmogony of the Bible, are essentially those brought out by Professor Guyot in his lectures.

Stages of Progress

These stages of progress are as follow:

1. The *Beginning of Activity in Matter.*—In such a beginning from matter in the state of a gaseous fluid the activity would be intense, and it would show itself at once by a manifestation of light, since light is a resultant of molecular activity. A flash of light through the universe would therefore be the first announcement of the work begun.
2. The development of the *Earth.*—A dividing and subdividing of the original fluid going on would have evolved systems of various grades, and ultimately the orbs of space, among these the earth, an igneous sphere enveloped in vapors.
3. The production of the *Earth's Physical Features,*—by the outlining of the continents and oceans. The condensible vapors would have gradually settled upon the earth as cooling progressed.
4. The introduction of *Life* under its simplest forms,—as in the lowest of plants, and perhaps, also, of animals. . . . As plants are primarily the food of animals, there is reason for believing that the idea of life was first expressed in a plant.
5. The display of the *Systems* in the Kingdoms of Life,—the exhibition of the four grand types under the Animal kingdom, being the predominant idea in this phase of progress.
6. The introduction of the highest class of Vertebrates—that of

the *Mammals* (the class to which *Man* belongs),—viviparous species, which are eminent above all other Vertebrates for a quality prophetic of a high moral purpose—that of suckling their young.
7. The introduction of *Man*,—the first being of moral and intellectual qualities, and one in whom the unity of nature has its full expression.

There is another great event in the Earth's history which has not yet been mentioned, because of a little uncertainty with regard to its exact place among the others. The event referred to is the first shining of the sun upon the earth, after the vapors which till then had shrouded the sphere were mostly condensed. This must have preceded the introduction of the Animal system, since the sun is the grand source of activity throughout nature on the earth, and is essential to the existence of life, excepting its lowest forms. . . .

The order will, then, be—

1. Activity begun—light an immediate result.
2. The earth made an independent sphere.
3. Outlining of the land and water, determining the earth's general configuration.
4. The idea of life expressed in the lowest plants, and afterwards, if not contemporaneously, in the lowest or systemless animals, or Protozoans.
5. The energizing light of the sun shining on the earth—an essential preliminary to the display of the systems of life.
6. Introduction of the systems of life.
7. Introduction of Mammals—the highest order of Vertebrates,—the class afterwards to be dignified by including a being of moral and intellectual nature.
8. Introduction of Man.

Cosmogony of the Bible

There is one ancient document on cosmogony—that of the opening page of the Bible—which is not only admired for its sublimity, but is very generally believed to be of divine origin, and which, therefore, demands at least a brief consideration in this place.

In the first place, it may be observed that *this document, if true, is of divine origin.* For no human mind was witness of the events; and no such mind in the early age of the world, unless gifted with superhuman intelligence, could have contrived such a scheme; would

The Character of Science

have placed the creation of the sun, the source of light to the earth, so long after the creation of light, even on the *fourth* day, and, what is equally singular, between the creation of plants and that of animals, when so important to both; and none could have reached to the depths of philosophy exhibited in the whole plan.

Again, *If divine, the account must bear marks of human imperfection, since it was communicated through man.* Ideas suggested to a human mind by the Deity would take shape in that mind according to its range of knowledge, modes of thought, and use of language, unless it were at the same time supernaturally gifted with the profound knowledge and wisdom adequate to their conception; and even then they could not be intelligibly expressed, for want of words to represent them.

The central thought of each step in the Scripture cosmogony—for example, Light; the dividing of the fluid earth from the fluid around it, individualizing the earth; the arrangement of its land and water; vegetation; and so on—is brought out in the simple and natural style of a sublime intellect, wise for its times, but unversed in the depths of science which the future was to reveal. The idea of vegetation to such a one would be vegetation as he knew it; and so it is described. The idea of dividing the earth from the fluid around it would take the form of a dividing from the fluid above, in the imperfect conceptions of a mind unacquainted with the earth's sphericity and the true nature of the firmament—especially as the event was beyond the reach of all ordinary thought.

Objections are often made to the word "day," as if its use limited the time of each of the six periods to a day of twenty-four hours. But in the course of the document this word "day" has various significations, and, among them, all that are common to it in ordinary language. These are: (1) The light—"God called the light day," v. 5; (2) the "evening and the morning" before the appearance of the sun; (3) the "evening and the morning" after the appearance of the sun; (4) the hours of light in the twenty-four hours (as well as the whole twenty-four hours), in verse 14; and (5) in the following chapter, at the commencement of another record of creation, the whole period of creation is called "a day." The proper meaning of "evening and morning," in a history of creation, is *beginning and completion;* and, in this sense, darkness before light is but a common metaphor.

A Deity working in creation like a day-laborer by earth-days of twenty-four hours, resting at night, is a belittling conception, and one probably never in the mind of the sacred penman. In the plan of an infinite God, centuries are required for the maturing of some of the plants with which the earth is adorned.

The order of events in the Scripture cosmogony corresponds essentially with that which has been given. There was first a void and formless earth: this was literally true of the "heavens and the earth," if they were in the condition of a gaseous fluid. The succession is as follows:

1. Light.
2. The dividing of the waters below from the waters above the earth (the word translated waters may mean fluid).
3. The dividing of the land and water on the earth.
4. Vegetation; which Moses, appreciating the philosophical characteristic of the new creation distinguishing it from previous inorganic substances, defines as that "which has seed in itself."
5. The sun, moon, and stars.
6. The lower animals, those that swarm in the waters, and the creeping and flying species of the land.
7. Beasts of prey ("creeping" here meaning "prowling").
8. Man.

In this succession, we observe not merely an order of events, like that deduced from science; there is a system in the arrangement, and a far-reaching prophecy, to which philosophy could not have attained, however instructed.

The account recognizes in creation two great eras of three days each—an *Inorganic* and an *Organic.*

Each of these eras opens with the appearance of *light:* the *first,* light cosmical; the *second,* light from the sun for the special uses of the earth.

Each era ends in a "day" of two great works—the two shown to be distinct by being severally pronounced "good." On the *third* "day," that closing the Inorganic era, there was first the *dividing of the land from the waters,* and afterwards the *creation of vegetation,* or the institution of a kingdom of life—a work widely diverse from all preceding it in the era. So on the *sixth* "day," terminating the Organic era, there was first *the creation of Mammals,* and then a second

The Character of Science

far greater work, totally new in its grandest element, *the creation of Man.*

The arrangement is, then, as follows:

1. The Inorganic Era.
   1st Day.—*Light* cosmical.
   2d Day.—The earth divided from the fluid around it, or individualized.
   3d Day.— { 1. Outlining of the land and water.
              2. Creation of vegetation.
2. The Organic Era.
   4th Day.—*Light* from the sun.
   5th Day.—Creation of the lower orders of animals.
   6th Day.— { 1. Creation of Mammals.
              2. Creation of Man. . . .

The record in the Bible is, therefore, profoundly philosophical in the scheme of creation which it presents. It is both true and divine. It is a declaration of authorship, both of Creation and the Bible, on the first page of the sacred volume.

There can be no real conflict between the two Books of the *Great Author*. Both are revelations made by Him to man—the *earlier* telling of God-made harmonies coming up from the deep past, and rising to their height when man appeared, the *later* teaching man's relations to his Maker, and speaking of loftier harmonies in the eternal future.

William Paley

PROOF OF GOD FROM THE WORLD OF BIOLOGY

William Paley's (1743–1805) lucid and classic summary of the argument for God from intricate biological organisms influenced generations of scientists, poets, educators, and clergymen. His argument below is drawn from his Natural Theology, *first published in 1802 and subject to numerous summaries, revisions, and elaborations. Like Charles Darwin, undergraduates in colleges in England and America were frequently required practically to memorize Paley's argument. Paley sought to prove that the contrivances of nature prove both the existence and personal characteristics (unity, intelligence, and benevolence) of the Deity, and the reading here indicates how his reasoning encouraged both religious awe and a "wonderfully curious" attachment to natural science. In the process, how much did Paley stimulate a romanticizing of nature?*

State of the Argument

In crossing a heath, suppose I pitched my foot against a *stone,* and were asked how the stone came to be there: I might possibly answer, that for anything I knew to the contrary, it had lain there forever: nor would it perhaps be very easy to show the absurdity of this answer. But suppose I had found a *watch* upon the ground, and it should be inquired how the watch happened to be in that place; I should hardly think of the answer which I had before given, that for anything I knew, the watch might have always been there. Yet why should not this answer serve for the watch as well as for the stone? Why is it not as admissible in the second case as in the first? For this reason, and for no other, *viz.* that when we come to inspect the watch, we perceive (what we could not discover in the stone) that its several parts are framed and put together for a purpose, e.g. that they are so formed and adjusted as to produce motion, and that motion so regulated as to point out the hour of the day; that, if the different parts had been differently shaped from what they are, of a different size from what they are, or placed after any other manner, or in any other order, than that in which they are placed, either no

---

Abridged from William Paley, *The Works of William Paley* (Philadelphia: J. J. Woodward, 1831).

motion at all would have been carried on in the machine, or none which would have answered the use that is now served by it. . . . This mechanism being observed (it requires indeed an examination of the instrument, and perhaps some previous knowledge of the subject, to perceive and understand it; but being once, as we have said, observed and understood), the inference, we think, is inevitable, that the watch must have had a maker; that there must have existed, at some time, and at some place or other, an artificer or artificers, who formed it for the purpose which we find it actually to answer; who comprehended its construction, and designed its use.

Suppose, in the next place, that the person who found the watch, should, after some time, discover that, in addition to all the properties which he had hitherto observed in it, it possessed the unexpected property of producing, in the course of its movement, another watch like itself (the thing is conceivable); that it contained within it a mechanism, a system of parts, a mould for instance, or a complex adjustment of lathes, files, and other tools, evidently and separately calculated for this purpose; let us inquire, what effect ought such a discovery to have upon his former conclusion.

1. The first effect would be to increase his admiration of the contrivance, and his conviction of the consummate skill of the contriver. Whether he regarded the object of the contrivance, the distinct apparatus, the intricate, yet in many parts intelligible mechanism, by which it was carried on, he would perceive, in this new observation, nothing but an additional reason for doing what he had already done —for referring the construction of the watch to design, and to supreme art. If that construction *without* this property, or, which is the same thing, before this property had been noticed, proved intention and art to have been employed about it; still more strong would the proof appear, when he came to the knowledge of this farther property, the crown and perfection of all the rest.

The conclusion which the *first* examination of the watch, of its works, construction, and movement, suggested, was, that it must have had, for the cause and author of that construction, an artificer, who understood its mechanism, and designed its use. This conclusion is invincible. A *second* examination presents us with a new discovery. The watch is found, in the course of its movement, to produce another watch, similar to itself; and not only so, but we perceive in it a system or organization, separately calculated for that purpose.

What effect would this discovery have or ought it to have, upon our former inference? What, as hath already been said, but to increase, beyond measure, our admiration of the skill which had been employed in the formation of such a machine? Or shall it, instead of this, all at once turn us round to an opposite conclusion, *viz.* that no art or skill whatever has been concerned in the business, although all other evidences of art and skill remain as they were, and this last and supreme piece of art be now added to the rest? Can this be maintained without absurdity? Yet this is atheism.

Application of the Argument

This is atheism: for every indication of contrivance, every manifestation of design, which existed in the watch, exists in the works of nature; with the difference, on the side of nature, of being greater and more, and that in a degree which exceeds all computation. I mean, that the contrivances of nature surpass the contrivances of art, in the complexity, subtlety, and curiosity, of the mechanism; and still more, if possible, do they go beyond them in number and variety; yet, in a multitude of cases, are not less evidently mechanical, not less evidently contrivances, not less evidently accommodated to their end, or suited to their office, than are the most perfect productions of human ingenuity.

I know no better method of introducing so large a subject, than that of comparing a single thing with a single thing; an eye, for example, with a telescope. As far as the examination of the instrument goes, there is precisely the same proof that the eye was made for vision, as there is that the telescope was made for assisting it. They are made upon the same principles; both being adjusted to the laws by which the transmission and refraction of rays of light are regulated. I speak not of the origin of the laws themselves; but such laws being fixed, the construction, in both cases, is adapted to them. For instance; these laws require, in order to produce the same effect, that the rays of light, in passing from water into the eye, should be refracted by a more convex surface, than when it passes out of air into the eye. Accordingly we find that the eye of a fish, in that part of it called the crystalline lens, is much rounder than the eye of terrestrial animals. What plainer manifestation of design can there be than this difference? What could a mathematical instrument-maker have done more, to show his knowledge of his principle, his application of that knowledge, his suiting of his means to his end; I will not say to display the compass or excellence of his skill and art, for in these all comparison is indecorous, but to testify counsel, choice, consideration, purpose?

But farther; there are other points, not so much perhaps of strict resemblance between the two, as of superiority of the eye over the telescope; yet of a superiority which, being founded in the laws that regulate both, may furnish topics of fair and just comparison. Two things were wanted to the eye, which were not wanted (at least in the same degree) to the telescope; and these were, the adaptation of the organ, first, to different degrees of light; and, secondly, to the vast diversity of distance at which objects are viewed by the naked eye, *viz.* from a few inches to as many miles.

But this, though much, is not the whole: by different species of animals the faculty we are describing is possessed, in degrees suited to the different range of vision which their mode of life, and of procuring their food requires. *Birds,* for instance, in general, procure their food by means of their beak; and, the distance between the eye and the point of the beak being small, it becomes necessary that they should have the power of seeing very near objects distinctly. On the other hand, from being often elevated much above the ground, living in air, and moving through it with great velocity, they require, for their safety, as well as for assisting them in descrying their prey, a power of seeing at a great distance; a power of which, in birds of rapine, surprising examples are given.

The eyes of *fishes* also, compared with those of terrestrial animals, exhibit certain distinctions of structure, adapted to their state and element. We have already observed upon the figure of the crystalline compensating it by its roundness the density of the medium through which their light passes. To which we have to add, that the eyes of fish, in their natural and indolent state, appear to be adjusted to near objects, in this respect differing from the human eye, as well as those of quadrupeds and birds. The ordinary shape of the fish's eye being in a much higher degree convex than that of land animals, a corresponding difference attends its muscular conformation, *viz.* that it is throughout calculated for *flattering* the eye.

The *iris* also in the eyes of fish does not admit of contraction. This is a great difference, of which the probable reason is, that the diminished light in water is never too strong for the retina.

Thus, in comparing the eyes of different kinds of animals, we see, in their resemblances and distinctions, one general plan laid down, and that plan varied with the varying exigencies to which it is to be applied.

It were, however, injustice to dismiss the eye as a piece of mechanism, without noticing that most exquisite of all contrivances, the *nictitating membrane,* which is found in the eyes of birds and of many quadrupeds. Its use is to sweep the eye, which it does in an instant, to spread over it the lachrymal humor; to defend it also from sudden injuries; yet not totally, when drawn upon the pupil, to shut out the light. The commodiousness with which it lies folded up in the upper corner of the eye, ready for use and action, and the quickness with which it executes its purpose, are properties known and obvious to every observer: but what is equally admirable, though not quite so obvious, is the combination of two kinds of substance, muscular and elastic, and of two different kinds of action, by which the motion of this membrane is performed.

One question may possibly have dwelt in the reader's mind during the perusal of these observations, namely, Why should not the Deity have given to the animal the faculty of vision *at once*? Why this circuitous perception; the ministry of so many means; an element provided for the purpose; reflected from opaque substances, refracted through transparent one; and both according to precise laws; then a complex organ, in intricate and artificial apparatus, in order, by the operation of this element, and in conformity with the restrictions of these laws, to produce an image upon a membrane communicating with the brain? Wherefore all this? Why make the difficulty in order to surmount it? If to perceive objects by some other mode than that of touch, or objects which lay out of the reach of that sense, were the thing proposed; could not a simple volition of the Creator have communicated the capacity? Why resort to contrivance, where power is omnipotent? Contrivance, by its very definition and nature, is the refuge of imperfection. To have recourse to expedients, implies difficulty, impediments, restraint, defect of power. This question belongs to the other senses, as well as to sight; to the general functions of animal life, as nutrition, secretion, respiration; to the

Proof of God from the World of Biology

economy of vegetables; and indeed to almost all the operations of nature. The question, therefore, is of very wide extent; and amongst other answers which may be given to it, besides reasons of which probably we are ignorant, one answer is this: It is only by the display of contrivance, that the existence, the agency, the wisdom, of the Deity, *could* be testified to his rational creatures. This is the scale by which we ascend to all the knowledge of our Creator which we possess so far as it depends upon the phenomena, or the works of nature. Take away this, and you take away from us every subject of observation, and ground of reasoning; I mean as our rational faculties are formed at present. Whatever is done, God could have done without the intervention of instruments or means; but it is in the construction of instruments, in the choice and adaptation of means, that a creative intelligence is seen. It is this which constitutes the order and beauty of the universe.

Of the Personality of the Deity

Contrivance, if established, appears to me to prove everything which we wish to prove. Amongst other things, it proves the *personality* of the Deity, as distinguished from what is sometimes called nature, sometimes called a principle: which terms, in the mouths of those who use them philosophically, seem to be intended, to admit and to express an efficacy, but to exclude and to deny a personal agent. Now that which can contrive, which can design, must be a person. These capacities constitute personality, for they imply consciousness and thought. They require that which can perceive an end or purpose; as well as the power of providing means, and of directing them to their end. They require a center in which perceptions unite, and from which volitions flow; which is mind. The acts of a mind prove the existence of a mind; and in whatever a mind resides, is a person. The seat of intellect is a person. We have no authority to limit the properties of mind to any corporeal form, or to any particular circumscription of space. These properties subsist, in created nature, under a great variety of sensible forms. Also every animated being has its sensorium; that is, a certain portion of space, within which perception and volition are exerted. This sphere may be enlarged to an indefinite extent; may comprehend the universe; and, being so imagined, may serve to furnish us with as good a notion, as we are capable of forming, of the *immensity* of the Divine Nature, i.e. of a Being, infinite, as well in essence as in power; yet nevertheless a person.

Wherever we see marks of contrivance, we are led for its cause to an *intelligent* author. And this transition of the understanding is founded upon uniform experience. We see intelligence constantly contriving; that is, we see intelligence constantly producing effects, marked and distinguished by certain properties; not certain particular properties, but by a kind and class of properties, such as relation to an end, relation of parts to one another, and to a common purpose. We see, wherever we are witnesses to the actual formation of things, nothing except intelligence producing effects so marked and distinguished. Furnished with this experience, we view the productions of nature. We observe *them* also marked and distinguished in the same manner. We wish to account for their origin. Our experience suggests a cause perfectly adequate to this account. *No* experience, no single instance or example, can be offered in favor of any other. In this cause therefore we ought to rest; in this cause the common sense of mankind has, in fact, rested, because it agrees with that, which, in all cases, is the foundation of knowledge—the undeviating course of their experience.

The marks of *design* are too strong to be gotten over. Design must have had a designer. That designer must have been a person. That person is GOD.

It is an immense conclusion, that there is a GOD; a perceiving, intelligent, designing Being; at the head of creation, and from whose will it proceeded. The *attributes* of such a Being, suppose his reality to be proved, must be adequate to the magnitude, extent, and multiplicity of his operations: which are not only vast beyond comparison with those performed by any other power; but, so far as respects our conceptions of them, infinite, because they are unlimited on all sides.

Yet the contemplation of a nature so exalted, however surely we arrive at the proof of its existence, overwhelms our faculties. The mind feels its powers sink under the subject.

* * *

Of the "Unity of the Deity," the proof is, the *uniformity* of plan observable in the universe. The universe itself is a system; each part either depending upon other parts, or being connected with other

Proof of God from the World of Biology

parts by some common law of action, or by the presence of some common substance.

The proof of the *divine goodness* rests upon two propositions: each, as we contend, capable of being made out by observations drawn from the appearances of nature.

The first is, "that, in a vast plurality of instances in which contrivance is perceived, the design of the contrivance is *beneficial.*"

The second, "that the Deity has superadded *pleasure* to animal sensations, beyond what was necessary for any other purpose, or when the purpose, so far as it was necessary, might have been effected by the operation of pain."

First, "In a vast plurality of instances in which contrivance is perceived, the design of the contrivance is *beneficial.*"

No productions of nature display contrivance so manifestly as the parts of animals; and the parts of animals have all of them, I believe, a real, and with very few exceptions, all of them a known and intelligible, subserviency to the use of the animal. Now, when the multitude of animals is considered, the number of parts in each, their figure and fitness, the faculties depending upon them, the variety of species, the complexity of structure, the success, in so many cases, and felicity of the result, we can never reflect, without the profoundest adoration, upon the character of that Being from whom all these things have proceeded: we cannot help acknowledging, what an exertion of benevolence creation was; of a benevolence how minute in its care, how vast in its comprehension!

It is a happy world after all. The air, the earth, the water, teem with delighted existence. In a spring noon, or a summer evening, on whichever side I turn my eyes, myriads of happy beings crowd upon my view. "The insect youth are on the wing." Swarms of new-born *flies* are trying their pinions in the air. Their sportive motions, their wanton mazes, their gratuitous activity, their continual change of place without use or purpose, testify their joy, and the exultation which they feel in their lately discovered faculties. A *bee* amongst the flowers in spring, is one of the most cheerful objects that can be looked upon. Its life appears to be all enjoyment; so busy, and so pleased; yet it is only a specimen of insect life, with which, by reason of the animal being half domesticated, we happen to be better acquainted than we are with that of others.

The *young* of all animals appear to me to receive pleasure simply from the exercise of their limbs and bodily faculties, without reference to any end to be attained, or any use to be answered by the exertion. A child, without knowing anything of the use of language, is in a high degree delighted with being able to speak.

But it is not for youth alone that the great Parent of creation hath provided. Happiness is found with the purring cat, no less than with the playful kitten; in the armchair of dozing age, as well as in either the sprightliness of the dance or the animation of the chase. To novelty, to acuteness of sensation, to hope, to ardor of pursuit, succeeds, what is, in no inconsiderable degree, an equivalent for them all, "perception of ease."

I shall not, I believe, be contradicted when I say, that if one train of thinking be more desirable than another, it is that which regards the phenomena of nature with a constant reference to a supreme intelligent Author. To have made this the ruling, the habitual sentiment of our minds, is to have laid the foundation of everything which is religious. The world thenceforth becomes a temple, and life itself one continued act of adoration. The change is no less than this: that, whereas formerly God was seldom in our thoughts, we can now scarcely look upon anything without perceiving its relation to him. Every organized natural body, in the provisions which it contains for its sustentation and propagation, testifies a care, on the part of the Creator, expressly directed to these purposes. We are on all sides surrounded by such bodies; examined in their parts, wonderfully curious; compared with one another, no less wonderfully diversified. So that the mind, as well as the eye, may either expatiate in variety and multitude, or fix itself down to the investigation of particular divisions of the science.

William Cowper
HYMNS TO THE GOD OF NATURE

One of the most popular and influential of English poets in the transitional, pre-Romantic period in the last decades of the eighteenth century was William Cowper (1731–1800). Beginning with the publication of Olney Hymns *in 1779, Cowper taught the religious to find in nature the indisputable evidence of the existence and benevolence of God. Perhaps the most famous and frequently used of these hymns is the one given here.*

The six-book-long poem, The Task *(first published in 1782) is Cowper's most significant "secular" work. The selections below from Book VI, entitled "The Winter Walk at Noon," explore the multifaceted meaning of nature which is depicted as a milieu of healing and joy, reflecting God as its maker in spite of the sin and cruelty of mankind.*

Light Shining Out of Darkness

God moves in a mysterious way
  His wonders to perform;
He plants his footsteps in the sea,
  And rides upon the storm.

Deep in unfathomable mines
  Of never-failing skill,
He treasures up his bright designs,
  And works his sovereign will.

Ye fearful saints, fresh courage take,
  The clouds ye so much dread
Are big with mercy, and shall break
  In blessings on your head.

Judge not the Lord by feeble sense,
  But trust him for his grace;
Behind a frowning providence
  He hides a smiling face.

His purposes will ripen fast,
  Unfolding every hour;
The bud may have a bitter taste,
  But sweet will be the flower.

Abridged from William Cowper, *The Poetical Works of William Cowper* (London: Macmillan, 1893).

> Blind unbelief is sure to err,
> And scan his work in vain:
> God is his own interpreter,
> And He will make it plain.

The Winter Walk at Noon[1]

The Lord of all, Himself through all diffused,
Sustains and is the life of all that lives.
Nature is but a name for an effect
Whose cause is God. He feeds the secret fire
By which the mighty process is maintained,
Who sleeps not, is not weary; in whose sight
Slow-circling ages are as transient days;
Whose work is without labour; whose designs
No flaw deforms, no difficulty thwarts;
And whose beneficence no charge exhausts.

In nature, from the broad majestic oak
To the green blade that twinkles in the sun,
Prompts with remembrance of a present God.
His presence, who made all so fair, perceived,
Makes all still fairer. As with him no scene
Is dreary, so with him all seasons please.
Though winter had been none, had man been true,
And earth be punished for its tenant's sake,
Yet not in vengeance; as this smiling sky,
So soon succeeding such an angry night,
And these dissolving snows, and this clear stream
Recovering fast its liquid music, prove.

   Here unmolested, through whatever sign
The sun proceeds, I wander; neither mist,
Nor freezing sky nor sultry, checking me,
Nor stranger intermeddling with my joy.
Even in the spring and playtime of the year,
That calls the unwonted villager abroad
With all her little ones, a sportive train,
To gather kingcups in the yellow mead,
And prink their hair with daisies, or to pick
A cheap but wholesome salad from the brook,
These shades are all my own. The timorous hare,
Grown so familiar with her frequent guest,
Scarce shuns me; and the stockdove unalarmed
Sits cooing in the pine-tree, nor suspends

---

[1] Lines 221–230, 250–261, 295–320, 348–391.

His long love-ditty for my near approach.
Drawn from his refuge in some lonely elm
That age or injury has hollowed deep,
Where on his bed of wool and matted leaves
He has outslept the winter, ventures forth
To frisk awhile, and bask in the warm sun,
The squirrel, flippant, pert, and full of play.
He sees me, and at once, swift as a bird,
Ascends the neighbouring beech; there whisks his brush,
And perks his ears, and stamps and scolds aloud,
With all the prettiness of feigned alarm,
And anger insignificantly fierce.

Man scarce had risen, obedient to His call
Who formed him from the dust, his future grave,
When he was crowned as never king was since.
God set the diadem upon his head,
And angel choirs attended. Wondering stood
The new-made monarch, while before him passed,
All happy, and all perfect in their kind,
The creatures, summoned from their various haunts
To see their sovereign, and confess his sway.
Vast was his empire, absolute his power,
Or bounded only by a law whose force
'Twas his sublimest privilege to feel
And own, the law of universal love.
He ruled with meekness, they obeyed with joy;
No cruel purpose lurked within his heart,
And no distrust of his intent in theirs.
So Eden was a scene of harmless sport,
Where kindness on his part who ruled the whole
Begat a tranquil confidence in all,
And fear as yet was not, nor cause for fear.
But sin marred all; and the revolt of man,
That source of evils not exhausted yet,
Was punished with revolt of his from him.
Garden of God, how terrible the change
Thy groves and lawns then witnessed! Every heart,
Each animal of every name, conceived
A jealousy and an instinctive fear,
And, conscious of some danger, either fled
Precipitate the loathed abode of man,
Or growled defiance in such angry sort,
As taught him too to tremble in his turn.
Thus harmony and family accord
Were driven from Paradise; and in that hour
The seeds of cruelty, that since have swelled

To such gigantic and enormous growth,
Were sown in human nature's fruitful soil.
Hence date the persecution and the pain
That man inflicts on all inferior kinds,
Regardless of their plaints. To make him sport,
To gratify the frenzy of his wrath,
Or his base gluttony, are causes good
And just in his account, why bird and beast
Should suffer torture, and the streams be dyed
With blood of their inhabitants impaled.

William Wordsworth
NATURE AS MORAL GUIDE AND DIVINE PRESENCE

These selections from the poetry of William Wordsworth (1770–1850) capture much of the very essence of romanticism—an intuitive sense of the sublimity of simple things and of the enchantment and spiritual meaning inherent in the world of nature. Wordsworth's sense of the divine presence and moral dimensions of nature are displayed in the lines from "Tintern Abbey" and "The Excursion." "Tintern Abbey" was one of the poems included in Wordsworth's and Samuel Taylor Coleridge's Lyrical Ballads (1798), a virtual romantic manifesto; and "Excursion" represented Wordsworth's most ambitious later work. These selections, as well as Wordsworth's "The Redbreast Chasing the Butterfly" (1802), show how romantics pictured nature as gentle and idyllic, fully capable of supplying emotional pleasure and spiritual succor.

Lines Composed a Few Miles above Tintern Abbey[1]

I came among these hills; when like a roe
I bounded o'er the mountains, by the sides
Of the deep rivers, and the lonely streams,
Wherever nature led: more like a man
Flying from something that he dreads, than one
Who sought the thing he loved. For nature then
(The coarser pleasures of my boyish days,

From William Wordsworth, The Complete Poetical Works of William Wordsworth (London: Macmillan, 1891).

[1] Lines 67–111.

And their glad animal movements all gone
    by)
To me was all in all.—I cannot paint
What then I was. The sounding cataract
Haunted me like a passion; the tall rock,
The mountain, and the deep and gloomy
    wood,
Their colors and their forms, were then to
    me
An appetite; a feeling and a love,
That had no need of a remoter charm,
By thought supplied, nor any interest
Unborrowed from the eye.—That time is
    past,
And all its aching joys are now no more,
And all its dizzy raptures. Not for this
Faint I, nor mourn nor murmur; other gifts
Have followed; for such loss, I would believe,
Abundant recompence. For I have learned
To look on nature, not as in the hour
Of thoughtless youth; but hearing oftentimes
The still, sad music of humanity,
Nor harsh nor grating, though of ample power
To chasten and subdue. And I have felt
A presence that disturbs me with the joy
Of elevated thoughts; a sense sublime
Of something far more deeply interfused,
Whose dwelling is the light of setting suns,
And the round ocean and the living air,
And the blue sky, and in the mind of man;
A motion and a spirit, that impels
All thinking things, all objects of all thought,
And rolls through all things. Therefore am I still
A lover of the meadows and the woods,
And mountains; and of all that we behold
From this green earth; of all the mighty world
Of eye, and ear,—both what they half create,
And what perceive; well pleased to recognize
In nature and the language of the sense,
The anchor of my purest thoughts, the nurse,
The guide, the guardian of my heart, and soul
Of all my moral being.

The Redbreast Chasing the Butterfly

Art thou the bird whom Man loves best,
The pious bird with the scarlet breast,

Our little English Robin;
The bird that comes about our doors
When Autumn-winds are sobbing?
Art thou the Peter of Norway Boors?
   Their Thomas in Finland,
   And Russia far inland?
The bird, that by some name or other
All men who know thee call their brother,
The darling of children and men?
Could Father Adam open his eyes
And see this sight beneath the skies,
He'd wish to close them again.
—If the Butterfly knew but his friend,
Hither his flight he would bend;
And find his way to me,
Under the branches of the tree:
In and out, he darts about;
Can this be the bird, to man so good,
That, after their bewildering,
Covered with leaves the little children,
   So painfully in the wood?
What ailed thee, Robin, that thou could'st pursue
   A beautiful creature,
That is gentle by nature?
Beneath the summer sky
From flower to flower let him fly;
'Tis all that he wishes to do.
The cheerer Thou of our in-door sadness,
He is the friend of our summer gladness:
What hinders, then, that ye should be
Playmates in the sunny weather,
And fly about in the air together!
His beautiful wings in crimson are drest,
A crimson as bright as thine own:
Would'st thou be happy in thy nest,
O pious Bird! whom man loves best,
Love him, or leave him alone!

The Excursion[2]

                But descending
From these imaginative heights, that yield
Far-stretching views into eternity,
Acknowledge that to Nature's humbler power
Your cherished sullenness is forced to bend

---

[2] From Book IV, "Despondency Corrected," lines 1187–1234.

Nature as Moral Guide and Divine Presence

Even here, where her amenities are sown
With sparing hand. Then trust yourself abroad
To range her blooming bowers, and spacious fields,
Where on the labors of the happy throng
She smiles, including in her wide embrace
City, and town, and tower,—and sea with ships
Sprinkled;—be our Companion while we track
Her rivers populous with gliding life;
While, free as air, o'er printless sands we march,
Or pierce the gloom of her majestic woods;
Roaming, or resting under grateful shade
In peace and meditative cheerfulness;
Where living things, and things inanimate,
Do speak, at Heaven's command, to eye and ear,
And speak to social reason's inner sense,
With inarticulate language.
                            For, the Man—
Who, in this spirit, communes with the
    Forms
Of nature, who with understanding heart
Both knows and loves such objects as excite
No morbid passions, no disquietude,
No vengeance, and no hatred—needs must feel
The joy of that pure principle of love
So deeply, that, unsatisfied with aught
Less pure and exquisite, he cannot choose
But seek for objects of a kindred love
In fellow-natures and a kindred joy.
Accordingly he by degrees perceives
His feelings of aversion softened down;
A holy tenderness pervade his frame.
His sanity of reason not impaired,
Say rather, all his thoughts now flowing clear,
From a clear fountain flowing, he looks round
And seeks for good; and finds the good he seeks:
Until abhorrence and contempt are things
He only knows by name; and, if he hear,
From other mouths, the language which they speak,
He is compassionate; and has no thought,
No feeling, which can overcome his love.

And further; by contemplating these
    Forms
In the relations which they bear to man,
He shall discern, how, through the various means
Which silently they yield, are multiplied
The spiritual presences of absent things.

Psalms 8:1, 3–9; Romans 1:18–20, 28–32
MANKIND'S PROXIMITY TO THE ANGELS

These biblical texts contain several crucial insights into the ways the West legitimated mankind's significance before 1859 (and indeed the way those who opposed or revised Darwin continued to legitimate mankind's uniqueness). In the face of man's inconsequential size as compared to the heavens, the Psalmist uses several themes to underscore human dignity. The phrase that man was created a "little less than God" was translated in pre-twentieth-century versions of the English Bible as made "but a little lower than the angels." Critics of Darwin set the latter phrase over against the belief that Darwin taught that man was closer to apes than angels. The texts from the book of Romans are representative of much that both the Old and New Testaments presuppose about mankind's moral (or immoral) image. In spite of this moral corruption, however, the biblical literature regarded man as unquestionably significant. Some of the reasons for that significance are offered here, not the least of which is the unarticulated presupposition that mankind is the continual object of divine concern.

Psalms 8

¹O Lord, our Lord,
how majestic is thy name in all the earth!

. . .

³When I look at thy heavens, the work of thy fingers,
    the moon and the stars which thou hast established;
⁴what is man that thou art mindful of him,
    and the son of man that thou dost care for him?
⁵Yet thou hast made him little less than God,
    and dost crown him with glory and honor.
⁶Thou hast given him dominion over the works of thy hands;
    thou hast put all things under his feet,
⁷all sheep and oxen,
    and also the beasts of the field,
⁸the birds of the air, and the fish of the sea,
    whatever passes along the paths of the sea.

⁹O Lord, our Lord,
how majestic is thy name in all the earth!

From the Revised Standard Version Bible, copyright 1946, 1952 and © 1971, and used by permission.

Mankind's Proximity to the Angels

Romans 1

18 For the wrath of God is revealed from heaven against all ungodliness and wickedness of men who by their wickedness suppress the truth. [19]For what can be known about God is plain to them, because God has shown it to them. [20]Ever since the creation of the world his invisible nature, namely, his eternal power and deity, has been clearly perceived in the things that have been made. So they are without excuse. . . .

28 And since they did not see fit to acknowledge God, God gave them up to a base mind and to improper conduct. [29]They were filled with all manner of wickedness, evil, covetousness, malice. Full of envy, murder, strife, deceit, malignity, they are gossips, [30]slanderers, haters of God, insolent, haughty, boastful, inventors of evil, disobedient to parents, [31]foolish, faithless, heartless, ruthless. [32]Though they know God's decree that those who do such things deserve to die, they not only do them but approve those who practice them.

William Kirby
THE NOBLEST OF SPECIES

William Kirby (1759–1850) was one of the eight highly respected English scientists chosen by the Royal Society to display the "power, wisdom and goodness of God" in the Bridgewater Treatises *(1833–1840) on subjects in natural history. Kirby's assignment was to study in detail the animal kingdom, a study that led him ultimately to devote two chapters to mammals and man, chapters from which the excerpts here are taken. In the light of his assignment, Kirby's study becomes an ideal model illustrating pre-Darwinian conceptions of man in relation to both the biblical world-view and the new data from natural history. Note how Kirby draws upon biblical themes to underscore man's significance and how he enthusiastically accepts the classification by Georges Cuvier, which placed man in an order of his own—as opposed to the classifications of Linnaeus, which according to Kirby "degraded" man by linking him too closely with monkeys.*

Mammalians

We are now arrived at the last and highest Class of the Animal Kingdom, to which man himself belongs, and of which he forms the summit: but though he may be said to belong to it in some respects, in others he stands aloof from it, as an insulated animal, and one exalted far above it, being created rather to govern its members, than to be the associate of the highest of them.

This Class includes many animals which are of the greatest utility to man, and without which he could scarcely exist, at least not in comfort; and others again that attack him and his property; and though the fear of him, in some degree, still remains upon them, also often excite that passion in his breast. But he of all animals is the only one, that by the exercise of his reasoning powers and faculties, can arm himself with factitious weapons enabling him to cope with the superior strength, the fierceness, claws, and teeth of the tiger or the lion, and to lay them dead at his feet when in the very act of springing upon him.

The animals of this Class, that are *terrestrial,* are all *quadrupeds,* and are mostly covered with fur or hair, longer or shorter, though in some, these hairs become quills, as in the porcupine, or spines, as in the hedgehog; others, like the serpents and lizards, are protected

Abridged from William Kirby, *On the History, Habits and Instincts of Animals,* II (London: William Pickering, 1835).

by scales, as the *Manis*; and some are incased in a hard coat of armor, often consisting of pieces so united as to form a kind of mosaic, as the armadillo, the *Chlamyphorus,* and probably the *Megatherium.*

In the *aquatic* Mammalians the legs are, more or less, converted into *fins,* or means of natation. The whole body constituting the Class, though sometimes varying in the manner, are all distinguished by *giving suck* to their young, on which account they were denominated by the Swedish naturalist [Linnaeus], *Mammalians.*

* * *

Order 8. Linnaeus . . . degraded *man* when he placed him in the same Order with the *monkey,* and even considered his genus *Homo* as consisting of two species, advancing the Ouran Outan to the honor of being his congener, and a second species of man. Cuvier has, with great propriety, separated man, the heir of immortality, and *whose spirit goeth upward,* from the beast that perisheth, and *whose spirit goeth downward,* and placed them in different Orders. Man has employed some animals in almost every Order, or taken them under his care; but there is only a single instance of a Quadrumane being so used. There is a kind of monkey, a native of Madagascar, which, being of a gentle disposition, the natives of the southern part of that island take when they are young, and educate, as we do hounds, for the chase.

The principal function of these animals is to live and move in the trees, amongst the branches in tropical countries, and they subsist upon fruits, roots, the eggs of birds, and insects. One object of their creation seems to be to hold the mirror to man, that he may see how ugly and disgusting an object he becomes when he gives himself up to vice and the slave of his passions. In fact, in every department of the animal kingdom, the moral instruction of his reasonable creature seems to have been one of the objects of Creative Wisdom: and the sloth and the glutton may be added to the mandril and baboon as equally calculated to cause him to view vice with disgust and abhorrence; as the bee, the ant, and the beaver, to excite him to industry, and prudence, and foresight; or the dove to peace and mutual love.

Man

After traversing the whole Animal Kingdom from its very lowest grades, and having arrived at Man, who confessedly stands at the head, and is the only *visible* king and lord of all the rest, it will be expected that I should devote a few pages to the world's master.

I shall consider him both physically and metaphysically; physically, as to his actual *position,* and as to his *action* upon his subjects and property, whether vegetable or animal; and metaphysically as to his connection with that world, to which his mind or spirit belongs. When I say that Man stands at the *head* of the creation, I do not mean to affirm that he combines in himself every physical attribute in perfection that is found in all the animals below him; for it is manifest to every one, that many of them far exceed him in the perfection of many of their organs, and in their qualities of various kinds. For *sight,* he cannot compete with the *eagle;* for *scent,* with the *hound,* or the *shark;* for *swiftness,* with the *roe-buck;* for strength and bulk, with the *elephant:* but it is in his *mind* that his superiority lies. There is in him a *spirit,* an immaterial substance which constitutes him the sole representative here on earth, of the *spirit of spirits.* He is the only member of the Animal Kingdom that partakes both of a heavenly and of an earthly nature—that belongs both to a material and an immaterial world: and on this account it was that God, when he had created man, constituted him king over the whole sphere of animals with which he had peopled this globe that we inhabit. When his unhappy *fall* took place, the Divine Image was impaired, and consequently the dominion over those creatures, which formed a part of it, was proportionably weakened, and reduced to its present standard. But still, though weakened, it is not abrogated; his subjects have not universally broken the yoke and burst the bonds of his dominion—a large portion of them still acknowledge him as their king and master; and those that he has not subdued so as to make them do his bidding, still fear him and flee him: and even of these, there is none so fierce and intractable, that he has not found means to tame and subdue. And this is the position in which he now stands with respect to the animal kingdom; he has that within him that enables him to master them, and apply such of them as are of a convertible nature, if I may so speak, to work his will and answer his purpose.

The action of man upon the world without him, is *threefold.* His *first* action upon them is, that of the mind to contemplate them, so as to gain a knowledge of their forms and structure—of their habits and instincts—of their meaning and uses. His *second* action upon

The Noblest of Species

them, having studied their natures, and discovered how they may be made profitable to him, is to collect and multiply such species as he finds will, in any way, answer his purpose. His *third* action upon them is to diminish and keep within due limits those species that experience teaches him are noxious and prejudicial either to himself, or those animals that he has taken into alliance with him, which are principal sources of wealth to him, and minister to his daily use, comfort, and enjoyment.

The highest view that we can take of man is that which looks upon him as belonging to a spiritual as well as a material world. . . . The beasts of the field honor him, and all creatures that he hath made glorify him. But man must study the book open before him; and the more he studies it, the more audible to him will be the general voice to his spiritual ear, and he will clearly perceive that every created thing glorifies God in its place, by fulfilling his will, and the great purpose of his providence; but that he himself alone can give a tongue to every creature, and pronounce for all a general doxology.

But further, in contemplating them, he will not only behold the glory of the Godhead reflected, but, from their several instincts and characters, he may derive much spiritual instruction.

In this enumeration and history of the principal tribes of the Animal Kingdom, we have traced in every page the footsteps of infinite Wisdom, Power, and Goodness. In our ascent from the most minute and least animated parts of that Kingdom to man himself, we have seen in every department that nothing was left to chance, or the rule of circumstances, but everything was adapted by its structure and organization for the situation in which it was to be placed, and the functions it was to discharge; that though every being, or group of beings, had separate interests, and wants, all were made to subserve to a common purpose, and to promote a common object; and that though there was a general and unceasing conflict between the members of this sphere of beings, introducing apparently death and destruction into every part of it, yet that by this great mass of seeming evil pervading the whole circuit of the animal creation, the renewed health and vigour of the entire system was maintained. A part suffers for the benefit and salvation of the whole; so that the doctrine of the sufferings of one creature, by the will of God, being necessary to promote the welfare of another, is irrefragably established by everything we see in nature; and further, that there is an unseen hand directing all to accomplish this great object, and taking care that the destruction shall in no case exceed the necessity. Well then, may all finally exclaim, in the words of the Divine Psalmist:—

O Lord, how manifold are thy works, in WISDOM hast thou made them all; the earth is full of thy riches.

II EVOLUTION, NATURE AND RELIGION

Charles Darwin
A NEW, REVOLUTIONARY WORLD VIEW

These readings group insights of Charles Darwin (1809–1882) under three headings. The first, Evolutionary Change versus a Fixed Creation, *deals with his theory of evolution by natural selection and shows what he meant by that theory and how he set it forcefully against the traditional idea of creation—including the theory of catastrophic eras of creation. Intentionally and directly he utilized his scientific opinions to create an intellectual revolution.*

The second set of readings, Evolution Undermines Teleology and Theism, *shows how Darwin rejected the Paleyean argument from design in nature and eventually became an agnostic. Excerpts from three of Darwin's writings are included: (1) "Organs of Extreme Perfection" is taken from the discussion in* The Origin of Species, *where Darwin expressly sought to invalidate William Paley's argument for God from the principle of design in nature; (2) the reading entitled here "Concerning Variabilities in Nature" is taken from the concluding discussion in* The Variation of Animals and Plants Under Domestication *(1868), in which Darwin rejected the idea (put forth by Asa Gray below) that divine providence oversees variations in the evolving organic world; (3) excerpts from Darwin's* Autobiography *indicate how and why Darwin's religious beliefs were transformed from orthodox Anglicanism to agnosticism.*

The third set of readings, Nature as Blundering, Low, and Horribly Cruel, *vividly portrays the new understanding of nature that was uncovered by Darwin, an understanding that lay at the foundation of his theory of evolution.*

EVOLUTIONARY CHANGE VERSUS A FIXED CREATION

When on board H. M. S. *Beagle* as naturalist, I was much struck with certain facts in the distribution of the inhabitants of South America, and in the geological relations of the present to the past inhabitants of that continent. These facts seemed to me to throw some light on the origin of species—that mystery of mysteries, as it has been called by one of our greatest philosophers. On my return home, it occurred to me, in 1837, that something might perhaps be made out on this question by patiently accumulating and reflecting on all sorts of facts which could possibly have any bearing on it. After five years'

Abridged from Charles Darwin, *On the Origin of Species* (1st ed.; London: John Murray, 1859).

work I allowed myself to speculate on the subject, and drew up some short notes; these I enlarged in 1844 into a sketch of the conclusions, which then seemed to me probable: from that period to the present day I have steadily pursued the same object. I hope that I may be excused for entering on these personal details, as I give them to show that I have not been hasty in coming to a decision.

In considering the origin of species, it is quite conceivable that a naturalist, reflecting on the mutual affinities of organic beings, on their embryological relations, their geographical distribution, geological succession, and other such facts, might come to the conclusion that each species had not been independently created, but had descended, like varieties, from other species. Nevertheless such a conclusion, even if well founded, would be unsatisfactory, until it could be shown how the innumerable species inhabiting this world have been modified, so as to acquire that perfection of structure and coadaptation which most justly excites our admiration.

It is, therefore, of the highest importance to gain a clear insight into the means of modification and coadaptation. At the commencement of my observations it seemed to me probable that a careful study of domesticated animals and of cultivated plants would offer the best chance of making out this obscure problem. Nor have I been disappointed; in this and in all other perplexing cases I have invariably found that our knowledge, imperfect though it be, of variation under domestication, afforded the best and safest clue. I may venture to express my conviction of the high value of such studies, although they have been very commonly neglected by naturalists.

From these considerations, I shall devote the first chapter of this Abstract to Variation under Domestication. We shall thus see that a large amount of hereditary modification is at least possible; and, what is equally or more important, we shall see how great is the power of man in accumulating by his Selection successive slight variations. I will then pass on to the variability of species in a state of nature; but I shall, unfortunately, be compelled to treat this subject far too briefly, as it can be treated properly only by giving long catalogues of facts. We shall, however, be enabled to discuss what circum-

---

FIGURE 2. An aging Darwin sits in his favorite chair and chuckles while others battle over his theories. (*Vanity Fair*; photograph by Richard V. T. Stearns)

stances are most favorable to variation. In the next chapter the Struggle for Existence amongst all organic beings throughout the world, which inevitably follows from their high geometrical powers of increase, will be treated of. This is the doctrine of Malthus, applied to the whole animal and vegetable kingdoms. As many more individuals of each species are born than can possibly survive; and as, consequently, there is a frequently recurring struggle for existence, it follows that any being, if it vary however slightly in any manner profitable to itself, under the complex and sometimes varying conditions of life, will have a better chance of surviving, and thus be *naturally selected.* From the strong principle of inheritance, any selected variety will tend to propagate its new and modified form.

No one ought to feel surprise at much remaining as yet unexplained in regard to the origin of species and varieties, if he makes due allowance for our profound ignorance in regard to the mutual relations of all the beings which live around us. Who can explain why one species ranges widely and is very numerous, and why another allied species has a narrow range and is rare? Yet these relations are of the highest importance, for they determine the present welfare, and, as I believe, the future success and modification of every inhabitant of this world. Still less do we know of the mutual relations of the innumerable inhabitants of the world during the many past geological epochs in its history. Although much remains obscure, and will long remain obscure, I can entertain no doubt, after the most deliberate study and dispassionate judgment of which I am capable, that the view which most naturalists entertain, and which I formerly entertained—namely, that each species has been independently created—is erroneous. I am fully convinced that species are not immutable; but that those belonging to what are called the same genera are lineal descendants of some other and generally extinct species, in the same manner as the acknowledged varieties of any one species are the descendants of that species. Furthermore, I am convinced that Natural Selection has been the main but not exclusive means of modification.

* * *

It may be asked, how is it that varieties, which I have called incipient species, become ultimately converted into good and distinct species, which in most cases obviously differ from each other

A New, Revolutionary World View

far more than do the varieties of the same species? How do those groups of species, which constitute what are called distinct genera, and which differ from each other more than do the species of the same genus, arise? All these results, as we shall more fully see in the next chapter, follow inevitably from the struggle for life. Owing to this struggle for life, any variation, however slight and from whatever cause proceeding, if it be in any degree profitable to an individual of any species, in its infinitely complex relations to other organic beings and to external nature, will tend to the preservation of that individual, and will generally be inherited by its offspring. The offspring, also, will thus have a better chance of surviving, for, of the many individuals of any species which are periodically born, but a small number can survive. I have called this principle, by which each slight variation, if useful, is preserved, by the term of Natural Selection, in order to mark its relation to man's power of selection. We have seen that man by selection can certainly produce great results, and can adapt organic beings to his own uses, through the accumulation of slight but useful variations, given to him by the hand of Nature. But Natural Selection, as we shall hereafter see, is a power incessantly ready for action, and is as immeasurably superior to man's feeble efforts, as the works of Nature are to those of Art.

A struggle for existence inevitably follows from the high rate at which all organic beings tend to increase. Every being, which during its natural lifetime produces several eggs or seeds, must suffer destruction during some period of its life, and during some season or occasional year, otherwise, on the principle of geometrical increase, its numbers would quickly become so inordinately great that no country could support the product. Hence as more individuals are produced than can possibly survive, there must in every case be a struggle for existence, either one individual with another of the same species, or with the individuals of distinct species, or with the physical conditions of life. It is the doctrine of Malthus applied with manifold force to the whole animal and vegetable kingdoms; for in this case there can be no artificial increase of food, and no prudential restraint from marriage. Although some species may be now increasing, more or less rapidly, in numbers, all cannot do so, for the world would not hold them.

* * *

Under domestication we see much variability. This seems to be mainly due to the reproductive system being eminently susceptible to changes in the conditions of life; so that this system, when not rendered impotent, fails to reproduce offspring exactly like the parent-form. Variability is governed by many complex laws—by correlation of growth, by use and disuse, and by the direct action of the physical conditions of life. There is much difficulty in ascertaining how much modification our domestic productions have undergone; but we may safely infer that the amount has been large, and that modifications can be inherited for long periods.

Man does not actually produce variability; he only unintentionally exposes organic beings to new conditions of life, and then nature acts on the organization, and causes variability. But man can and does select the variations given to him by nature, and thus accumulate them in any desired manner. He thus adapts animals and plants for his own benefit or pleasure. . . . This process of selection has been the great agency in the production of the most distinct and useful domestic breeds. That many of the breeds produced by man have to a large extent the character of natural species, is shown by the inextricable doubts whether very many of them are varieties or aboriginal species.

There is no obvious reason why the principles which have acted so efficiently under domestication should not have acted under nature. In the preservation of favored individuals and races, during the constantly recurrent Struggle for Existence, we see the most powerful and ever-acting means of selection. The struggle for existence inevitably follows from the high geometrical ratio of increase which is common to all organic beings. This high rate of increase is proved by calculation, by the effects of a succession of peculiar seasons, and by the results of naturalization, as explained in the third chapter. More individuals are born than can possibly survive. A grain in the balance will determine which individual shall live and which shall die—which variety or species shall increase in number, and which shall decrease, or finally become extinct. As the individuals of the same species come in all respects into the closest competition with each other, the struggle will generally be most severe between them; it will be almost equally severe between the varieties of the same species, and next in severity between the species of the same genus. But the struggle will often be very severe between beings most remote in the scale of nature. The slightest advantage in one being, at any age or during any season, over those with which it comes into competition, or better adaptation in however slight a degree to the surrounding physical conditions, will turn the balance.

If then we have under nature variability and a powerful agent always ready to act and select, why should we doubt that variations in any way useful to beings, under their excessively complex relations of life, would be preserved, accumulated, and inherited? Why, if man can by patience select variations most useful to himself, should nature fail in selecting variations useful, under changing conditions of life, to her living products? What limit can be put to this power, acting during long ages and rigidly scrutinizing the whole constitution, structure, and habits of each creature—favoring the good and rejecting the bad? I can see no limit to this power, in slowly and beautifully adapting each form to the most complex relations of life. The theory of natural selection, even if we look no further than this, seems to me to be in itself probable. . . . Now let us turn to the special acts and arguments in favor of the theory.

As each species tends by its geometrical ratio of reproduction to increase inordinately in number; and as the modified descendants of each species will be enabled to increase by so much the more as they become more diversified in habits and structure, so as to be enabled to seize on many and widely different places in the economy of nature, there will be a constant tendency in natural selection to preserve the most divergent offspring of any one species. Hence during a long-continued course of modification, the slight differences, characteristic of varieties of the same species, tend to be augmented into the greater differences characteristic of species of the same genus. New and improved varieties will inevitably supplant and exterminate the older, less improved and intermediate varieties; and thus species are rendered to a large extent defined and distinct objects. Dominant species belonging to the larger groups tend to give birth to new and dominant forms; so that each large group tends to become still larger, and at the same time more divergent in character. But as all groups cannot thus succeed in increasing in size, for the world would not hold them, the more dominant groups beat the less dominant. This tendency in the large groups to go on increasing in size and diverging in character, together with the almost inevitable contingency of much extinction, explains the arrangement of all the forms of life, in groups subordinate to groups, all within a few great classes, which we now see everywhere around us, and which has prevailed throughout all time. This grand fact of the grouping of all organic beings seems to me utterly inexplicable on the theory of creation.

As natural selection acts solely by accumulating slight, successive, favorable variations, it can produce no great or sudden modification; it can act only by very short and slow steps. Hence the canon of "Natura non facit saltum," which every fresh addition to our knowledge tends to make more strictly correct, is on this theory simply intelligible. We can plainly see why nature is prodigal in variety, though niggard in innovation. But why this should be a law of nature if each species has been independently created, no man can explain.

Many other facts are, as it seems to me, explicable on this theory. How strange it is that a bird, under the form of a woodpecker, should have been created to prey on insects on the ground; that upland geese, which never or rarely swim, should have been created with webbed feet; that a thrush should have been created to dive and feed on sub-aquatic insects; and that a petrel should have been created with habits and structure fitting it for the life of an auk or grebe! and so on in endless other cases. But on the view of each species constantly trying to increase in number, with natural selection always ready to adapt the slowly varying descendants of each to any unoccupied or ill-occupied place in nature, these facts cease to be strange, or perhaps might even have been anticipated.

As natural selection acts by competition, it adapts the inhabitants of each country only in relation to the degree of perfection of their associates; so that we need feel no surprise at the inhabitants of any one country, although on the ordinary view supposed to have been specially created and adapted for that country, being beaten and supplanted by the naturalized productions from another land. Nor ought we to marvel if all the contrivances in nature be not, as far as we can judge, absolutely perfect; and if some of them be abhorrent to our ideas of fitness. We need not marvel at the sting of the bee causing the bee's own death; at drones being produced in such vast numbers for one single act, and being then slaughtered by their sterile sisters; at the astonishing waste of pollen by our fir-trees; at the instinctive hatred of the queen bee for her own fertile daughters; at ichneumonidae feeding within the live bodies of caterpillars; and at other such cases. The wonder indeed is, on the theory of natural selection, that more cases of the want of absolute perfection have not been observed.

In both varieties and species, use and disuse seem to have produced some effect; for it is difficult to resist this conclusion when we look, for instance, at the logger-headed duck, which has wings, incapable of flight, in nearly the same condition as in the domestic duck; or when we look at the burrowing tucu-tucu, which is occasionally blind, and then at certain moles, which are habitually blind and have their eyes covered with skin; or when we look at the blind animals inhabiting the dark caves of America and Europe. In both varieties and species correlation of growth seems to have played a most important part, so that when one part has been modified other parts are necessarily modified. In both varieties and species reversions to long-lost characters occur. How inexplicable on the theory of creation is the occasional appearance of stripes on the shoulder and legs of the several species of the horse-genus and in their hybrids! How simply is this fact explained if we believe that these species have descended from a striped progenitor, in the same manner as the several domestic breeds of pigeon have descended from the blue and barred rock-pigeon!

On the ordinary view of each species having been independently created, why should the specific characters, or those by which the species of the same genus differ from each other, be more variable than the generic characters in which they all agree? Why, for instance, should the color of a flower be more likely to vary in any one species of a genus, if the other species, supposed to have been created independently, have differently colored flowers, than if all the species of the genus have the same-colored flowers? If species are only well-marked varieties, of which the characters have become in a high degree permanent, we can understand this fact; for they have already varied since they branched off from a common progenitor in certain characters, by which they have come to be specifically distinct from each other; and therefore these same characters would be more likely still to be variable than the generic characters which have been inherited without change for an enormous period. It is inexplicable on the theory of creation why a part developed in a very unusual manner in any one species of a genus, and therefore, as we may naturally infer, of great importance to the species, should be eminently liable to variation; but, on my view, this part has undergone, since the several species branched off from a common progenitor, an unusual amount of variability and modification, and therefore we might expect this part generally to be still variable. But a part may be developed in the most unusual manner, like the wing of a bat, and yet not to be more variable than any other structure, if the part be common to many subordinate forms, that is, if it has been inherited for a very long period; for in this case it will have been rendered constant by long-continued natural selection.

If species be only well-marked and permanent varieties, we can at once see why their crossed offspring should follow the same complex laws in their degrees and kinds of resemblance to their parents—in being absorbed into each other by successive crosses, and in other such points—as do the crossed offspring of acknowledged varieties. On the other hand, these would be strange facts if species have been independently created, and varieties have been produced by secondary laws.

Looking to geographical distribution, if we admit that there has been during the long course of ages much migration from one part of the world to another, owing to former climatal and geographical changes, and to the many occasional and unknown means of dispersal, then we can understand, on the theory of descent with modification, most of the great leading facts in Distribution. We can see why there should be so striking a parallelism in the distribution of organic beings throughout space, and in their geological succession throughout time; for in both cases the beings have been connected by the bond of ordinary generation, and the means of modification have been the same. We see the full meaning of the wonderful fact, which must have struck every traveller, namely, that on the same continent, under the most diverse conditions, under heat and cold, on mountain and lowland, on deserts and marshes, most of the inhabitants within each great class are plainly related; for they will generally be descendants of the same progenitors and early colonists. . . . Although two areas may present the same physical conditions of life, we need feel no surprise at their inhabitants being widely different, if they have been for a long period completely separated from each other; for as the relation of organism to organism is the most important of all relations, and as the two areas will have received colonists from some third source or from each other, at various periods and in different proportions, the course of modification in the two areas will inevitably be different.

On this view of migration, with subsequent modification, we can see why oceanic islands should be inhabited by few species, but of these, that many should be peculiar. We can clearly see why those animals which cannot cross wide spaces of ocean, as frogs and terrestrial mammals, should not inhabit oceanic islands; and why, on the other hand, new and peculiar species of bats, which can traverse the ocean, should so often be found on islands far distant from any continent. Such facts as the presence of peculiar species of bats, and the absence of all other mammals, on oceanic islands, are utterly inexplicable on the theory of independent acts of creation.

The fact, as we have seen, that all past and present organic beings constitute one grand natural system, with group subordinate to group, and with extinct groups often falling in between recent groups, is intelligible on the theory of natural selection with its contingencies of extinction and divergence of character. On these same principles we see how it is, that the mutual affinities of the species and genera within each class are so complex and circuitous. We see why certain characters are far more serviceable than others for classification—why adaptive characters, though of paramount importance to the being, are of hardly any importance in classification; why characters derived from rudimentary parts, though of no service to the being, are often of high classificatory value; and why embryological characters are the most valuable of all. The real affinities of all organic beings are due to inheritance or community of descent. The natural system is a genealogical arrangement, in which we have to discover the lines of descent by the most permanent characters, however slight their vital importance may be.

The framework of bones being the same in the hand of a man, wing of a bat, fin of the porpoise, and leg of the horse—the same number of vertebrae forming the neck of the giraffe and of the elephant—and innumerable other such facts, at once explain themselves on the theory of descent with slow and slight successive modifications.

Although I am fully convinced of the truth of the views given in this volume under the form of an abstract, I by no means expect to convince experienced naturalists whose minds are stocked with a multitude of facts all viewed, during a long course of years, from a point of view directly opposite to mine. It is so easy to hide our ignorance under such expressions as the "plan of creation," "unity of design," etc., and to think that we give an explanation when we only restate a fact. Anyone whose disposition leads him to attach more weight to unexplained difficulties than to the explanation of a certain number of facts will certainly reject my theory. A few naturalists, endowed with much flexibility of mind, and who have already begun to doubt on the immutability of species, may be influenced by this volume; but I look with confidence to the future, to young and rising naturalists, who will be able to view both sides of the question with impartiality. Whoever is led to believe that species are mutable will do good service by conscientiously expressing his conviction; for only thus can the load of prejudice by which this subject is overwhelmed be removed.

Several eminent naturalists have of late published their belief that a multitude of reputed species in each genus are not real species; but that other species are real, that is, have been independently created. This seems to me a strange conclusion to arrive at. They admit that a multitude of forms, which till lately they themselves thought were special creations, and which are still thus looked at by the majority of naturalists, and which consequently have every external characteristic feature of true species—they admit that these have been produced by variation, but they refuse to extend the same view to other and very slightly different forms. Nevertheless they do not pretend that they can define, or even conjecture, which are the created forms of life, and which are those produced by secondary laws. They admit variation as a *vera causa* in one case, they arbitrarily reject it in another, without assigning any distinction in the two cases. The day will come when this will be given as a curious illustration of the blindness of preconceived opinion. These authors seem no more startled at a miraculous act of creation than at an ordinary birth. But do they really believe that at innumerable periods in the earth's history certain elemental atoms have been commanded suddenly to flash into living tissues? Do they believe that at each supposed act of creation one individual or many were produced? Were all the infinitely numerous kinds of animals and plants created as eggs or seed, or as full grown? And in the case of mammals, were they created bearing the false marks of nourishment from the mother's womb? Although naturalists very properly demand a full explanation of every difficulty from those who believe in the mutability of species, on their own side they ignore the whole subject of the first appearance of species in what they consider reverent silence.

It may be asked how far I extend the doctrine of the modification of species. The question is difficult to answer, because the more distinct the forms are which we may consider, by so much the arguments fall away in force. But some arguments of the greatest weight extend very far. All the members of whole classes can be connected together by chains of affinities, and all can be classified on the same principle, in groups subordinate to groups. Fossil remains sometimes tend to fill up very wide intervals between existing orders. Organs in a rudimentary condition plainly show that an early progenitor had the organ in a fully developed state; and this in some instances necessarily implies an enormous amount of modification in the descendants. Throughout whole classes various structures are formed on the same pattern, and at an embryonic age the species closely resemble each other. Therefore I cannot doubt that the theory of descent with modification embraces all the members of the same class. I believe that animals have descended from at most only four or five progenitors, and plants from an equal or lesser number.

Analogy would lead me one step further, namely, to the belief that all animals and plants have descended from some one prototype. But analogy may be a deceitful guide. Nevertheless all living things have much in common, in their chemical composition, their germinal vesicles, their cellular structure, and their laws of growth and reproduction.

When the views entertained in this volume on the origin of species, or when analogous views are generally admitted, we can dimly foresee that there will be a considerable revolution in natural history. Systematists will be able to pursue their labors as at present; but they will not be incessantly haunted by the shadowy doubt whether this or that form be in essence a species. This I feel sure, and I speak after experience, will be no slight relief.

As species are produced and exterminated by slowly acting and still existing causes, and not by miraculous acts of creation and by catastrophes; and as the most important of all causes of organic change is one which is almost independent of altered and perhaps suddenly altered physical conditions, namely, the mutual relation of organism to organism—the improvement of one being entailing the improvement or the extermination of others; it follows, that the amount of organic change in the fossils of consecutive formations probably serves as a fair measure of the lapse of actual time. A number of species, however, keeping in a body might remain for a long period unchanged, whilst within this same period, several of these species, by migrating into new countries and coming into competition with foreign associates, might become modified; so that we must not overrate the accuracy of organic change as a measure of time. During early periods of the earth's history, when the forms of life were probably fewer and simpler, the rate of change was probably slower; and at the first dawn of life, when very few forms of the simplest structure existed, the rate of change may have been slow in an extreme degree. The whole history of the world, as at present known, although of a length quite incomprehensible by us, will hereafter be recognized as a mere fragment of time, compared with the ages which have elapsed since the first creature, the progenitor of innumerable extinct and living descendants, was created.

In the distant future I see open fields for far more important researches. Psychology will be based on a new foundation, that of the necessary acquirement of each mental power and capacity by gradation. Light will be thrown on the origin of man and his history.

Authors of the highest eminence seem to be fully satisfied with the view that each species has been independently created. To my mind it accords better with what we know of the laws impressed on matter by the Creator, that the production and extinction of the past and present inhabitants of the world should have been due to secondary causes, like those determining the birth and death of the individual. When I view all beings not as special creations, but as the lineal descendants of some few beings which lived long before the first bed of the Silurian system was deposited, they seem to me to become ennobled. Judging from the past, we may safely infer that not one living species will transmit its unaltered likeness to a distant futurity. And of the species now living very few will transmit progeny of any kind to a far distant futurity; for the manner in which all organic beings are grouped, shows that the greater number of species of each genus, and all the species of many genera, have left no descendants, but have become utterly extinct. We can so far take a prophetic glance into futurity as to foretell that it will be the common and widely spread species, belonging to the larger and dominant groups, which will ultimately prevail and procreate new and dominant species. As all the living forms of life are the lineal descendants of those which lived long before the Silurian epoch, we may feel certain that the ordinary succession by generation has never once been broken, and that no cataclysm has desolated the whole world. Hence we may look with some confidence to a secure future of equally inappreciable length. And as natural selection works solely by and for the good of each being, all corporeal and mental endowments will tend to progress towards perfection.

EVOLUTION UNDERMINES TELEOLOGY AND THEISM

Organs of Extreme Perfection

To suppose that the eye, with all its inimitable contrivances for adjusting the focus to different distances, for admitting different amounts of light, and for the correction of spherical and chromatic aberration, could have been formed by natural selection, seems, I freely confess, absurd in the highest possible degree. Yet reason tells me, that if numerous gradations from a perfect and complex eye to one very imperfect and simple, each grade being useful to its possessor, can be shown to exist; if, further, the eye does vary ever so slightly, and the variations be inherited, which is certainly the case; and if any variation or modification in the organ be ever useful to an animal under changing conditions of life, then the difficulty of believing that a perfect and complex eye could be formed by natural selection, though insuperable by our imagination, can hardly be considered real. How a nerve comes to be sensitive to light, hardly concerns us more than how life itself first originated; but I may remark that several facts make me suspect that any sensitive nerve may be rendered sensitive to light, and likewise to those coarser vibrations of the air which produce sound.

Abridged from Charles Darwin, *On the Origin of Species* (1st ed.; London: John Murray, 1859); Darwin, *The Variation of Animals and Plants Under Domestication*, II (New York: D. Appleton, 1883 [1868]); and Darwin, *The Life and Letters of Charles Darwin*, I, ed. by Francis Darwin (New York: D. Appleton, 1897).

In the Articulata we can commence a series with an optic nerve merely coated with pigment, and without any other mechanism; and from this low stage numerous gradations of structure, branching off in two fundamentally different lines, can be shown to exist, until we reach a moderately high stage of perfection. In certain crustaceans, for instance, there is a double cornea, the inner one divided into facets, within each of which there is a lens-shaped swelling. In other crustaceans the transparent cones which are coated by pigment, and which properly act only by excluding lateral pencils of light, are convex at their upper ends and must act by convergence; and at their lower ends there seems to be an imperfect vitreous substance. With these facts, here far too briefly and imperfectly given, which show that there is much graduated diversity in the eyes of living crustaceans, and bearing in mind how small the number of living animals is in proportion to those which have become extinct, I can see no very great difficulty (not more than in the case of many other structures) in believing that natural selection has converted the simple apparatus of an optic nerve merely coated with pigment and invested by transparent membrane, into an optical instrument as perfect as is possessed by any member of the great Articulate class.

He who will go thus far, if he finds on finishing this treatise that large bodies of facts, otherwise inexplicable, can be explained by the theory of descent, ought not to hesitate to go further, and to admit that a structure even as perfect as the eye of an eagle might be formed by natural selection, although in this case he does not know any of the transitional grades. His reason ought to conquer his imagination; though I have felt the difficulty far too keenly to be surprised at any degree of hesitation in extending the principle of natural selection to such startling lengths.

It is scarcely possible to avoid comparing the eye to a telescope. We know that this instrument has been perfected by the long-continued efforts of the highest human intellects; and we naturally infer that the eye has been formed by a somewhat analogous process. But may not this inference be presumptuous? Have we any right to assume that the Creator works by intellectual powers like those of man? If we must compare the eye to an optical instrument, we ought in imagination to take a thick layer of transparent tissue, with a nerve sensitive to light beneath, and then suppose every part of this layer to be continually changing slowly in density, so as to separate

A New, Revolutionary World View

into layers of different densities and thicknesses, placed at different distances from each other, and with the surfaces of each layer slowly changing in form. Further we must suppose that there is a power always intently watching each slight accidental alteration in the transparent layers; and carefully selecting each alteration which, under varied circumstances, may in any way, or in any degree, tend to produce a distincter image. We must suppose each new state of the instrument to be multiplied by the million; and each to be preserved till a better be produced, and then the old ones to be destroyed. In living bodies, variation will cause the slight alterations, generation will multiply them almost infinitely, and natural selection will pick out with unerring skill each improvement. Let this process go on for millions on millions of years; and during each year on millions of individuals of many kinds; and may we not believe that a living optical instrument might thus be formed as superior to one of glass, as the works of the Creator are to those of man?

If it could be demonstrated that any complex organ existed, which could not possibly have been formed by numerous, successive, slight modifications, my theory would absolutely break down. But I can find out no such case.

Although in many cases it is most difficult to conjecture by what transitions an organ could have arrived at its present state; yet, considering that the proportion of living and known forms to the extinct and unknown is very small, I have been astonished how rarely an organ can be named, towards which no transitional grade is known to lead. The truth of this remark is indeed shown by that old canon in natural history of "Natura non facit saltum." We meet with this admission in the writings of almost every experienced naturalist; or, as Milne Edwards has well expressed it, nature is prodigal in variety, but niggard in innovation. Why, on the theory of Creation, should this be so? Why should all the parts and organs of many independent beings, each supposed to have been separately created for its proper place in nature, be so invariably linked together by graduated steps? Why should not Nature have taken a leap from structure to structure? On the theory of natural selection, we can clearly understand why she should not; for natural selection can act only by taking advantage of slight successive variations; she can never take a leap, but must advance by the shortest and slowest steps.

Concerning Variabilities in Nature

An omniscient Creator must have foreseen every consequence which results from the laws imposed by Him. But can it be reasonably maintained that the Creator intentionally ordered, if we use the words in any ordinary sense, that certain fragments of rock should assume certain shapes so that the builder might erect his edifice? If the various laws which have determined the shape of each fragment were not predetermined for the builder's sake, can it be maintained with any greater probability that He specially ordained for the sake of the breeder each of the innumerable variations in our domestic animals and plants—many of these variations being of no service to man, and not beneficial, far more often injurious, to the creatures themselves? Did He ordain that the crop and tail-feathers of the pigeon should vary in order that the fancier might make his grotesque pouter and fantail breeds? Did He cause the frame and mental qualities of the dog to vary in order that a breed might be formed of indomitable ferocity, with jaws fitted to pin down the bull for man's brutal sport? But if we give up the principle in one case—if we do not admit that the variations of the primeval dog were intentionally guided in order that the greyhound, for instance, that perfect image of symmetry and vigor, might be formed—no shadow of reason can be assigned for the belief that variations, alike in nature and the result of the same general laws, which have been the groundwork through natural selection of the formation of the most perfectly adapted animals in the world, man included, were intentionally and specially guided. However much we may wish it, we can hardly follow Professor Asa Gray in his belief that "variation has been led along certain beneficial lines," like a stream "along definite and useful lines of irrigation." If we assume that each particular variation was from the beginning of all time preordained, then that plasticity of organization, which leads to many injurious deviations of structure, as well as the redundant power of reproduction which inevitably leads to a struggle for existence, and, as a consequence, to the natural selection or survival of the fittest, must appear to us superfluous laws of nature. On the other hand, an omnipotent and omniscient Creator ordains everything and foresees everything. Thus we are brought face to face with a difficulty as insoluble as is that of free will and predestination.

A New, Revolutionary World View

Autobiography [1876]

Cambridge 1828–1831.—After having spent two sessions in Edinburgh, my father perceived, or he heard from my sisters, that I did not like the thought of being a physician, so he proposed that I should become a clergyman. He was very properly vehement against my turning into an idle sporting man, which then seemed my probable destination. I asked for some time to consider, as from what little I had heard or thought on the subject I had scruples about declaring my belief in all the dogmas of the Church of England; though otherwise I liked the thought of being a country clergyman. Accordingly I read with care "Pearson on the Creed," and a few other books on divinity; and as I did not then in the least doubt the strict and literal truth of every word in the Bible, I soon persuaded myself that our Creed must be fully accepted.

Considering how fiercely I have been attacked by the orthodox, it seems ludicrous that I once intended to be a clergyman. Nor was this intention and my father's wish ever formally given up, but died a natural death when, on leaving Cambridge, I joined the *Beagle* as naturalist.

In order to pass the B.A. examination, it was also necessary to get up Paley's "Evidences of Christianity," and his "Moral Philosophy." This was done in a thorough manner, and I am convinced that I could have written out the whole of the "Evidences" with perfect correctness, but not of course in the clear language of Paley. The logic of this book and, as I may add, of his "Natural Theology," gave me as much delight as did Euclid. The careful study of these works, without attempting to learn any part by rote, was the only part of the academical course which, as I then felt and as I still believe, was of the least use to me in the education of my mind. I did not at that time trouble myself about Paley's premises; and taking these on trust, I was charmed and convinced by the long line of argumentation.

* * *

During these two years [Oct. 1836–Jan. 1839] I was led to think much about religion. Whilst on board the *Beagle* I was quite orthodox, and I remember being heartily laughed at by several of the officers (though themselves orthodox) for quoting the Bible as an unanswerable authority on some point of morality. I suppose it was the novelty of the argument that amused them. But I had gradually come by this time, i.e. 1836 to 1839, to see that the Old Testament was no more to be trusted than the sacred books of the Hindus. The question then continually rose before my mind and would not be banished—is it credible that if God were now to make a revelation to the Hindus, he would permit it to be connected with the belief in Vishnu, Siva, etc., as Christianity is connected with the Old Testament? This appeared to me utterly incredible.

By further reflecting that the clearest evidence would be requisite to make any sane man believe in the miracles by which Christianity is supported, and that the more we know of the fixed laws of nature the more incredible do miracles become; that the men at that time were ignorant and credulous to a degree almost incomprehensible by us; that the Gospels cannot be proved to have been written simultaneously with the events, that they differ in many important details, far too important, as it seemed to me, to be admitted as the usual inaccuracies of eye-witnesses—by such reflections as these, which I give not as having the least novelty or value, but as they influenced me, I gradually came to disbelieve in Christianity as a divine revelation. The fact that many false religions have spread over large portions of the earth like wild-fire had some weight with me.

But I was very unwilling to give up my belief; I feel sure of this, for I can well remember often and often inventing day-dreams of old letters between distinguished Romans, and manuscripts being discovered at Pompeii or elsewhere, which confirmed in the most striking manner all that was written in the Gospels. But I found it more and more difficult, with free scope given to my imagination, to invent evidence which would suffice to convince me. Thus disbelief crept over me at a very slow rate, but was at last complete. The rate was so slow that I felt no distress.

Although I did not think much about the existence of a personal God until a considerably later period of my life, I will here give the vague conclusions to which I have been driven. The old argument from design in Nature, as given by Paley, which formerly seemed to me so conclusive, fails, now that the law of natural selection has been discovered. We can no longer argue that, for instance, the beautiful hinge of a bivalve shell must have been made by an intelligent being, like the hinge of a door by man. There seems to be no more design in the variability of organic beings, and in the action of natural selection, than in the course which the wind blows. But I have discussed this subject at the end of my book on the "Variations of Domesticated Animals and Plants," and the argument there given has never, as far as I can see, been answered.

But passing over the endless beautiful adaptations which we everywhere meet with, it may be asked how can the generally beneficent arrangement of the world be accounted for? Some writers indeed are so much impressed with the amount of suffering in the world, that they doubt, if we look to all sentient beings, whether there is more of misery or of happiness; whether the world as a whole is a good or bad one. According to my judgment happiness decidedly prevails, though this would be very difficult to prove. If the truth of this conclusion be granted, it harmonizes well with the effects which we might expect from natural selection. If all the individuals of any species were habitually to suffer to an extreme degree, they would neglect to propagate their kind; but we have no reason to believe that this has ever, or at least often occurred. Some other considerations, moreover, lead to the belief that all sentient beings have been formed so as to enjoy, as a general rule, happiness.

Everyone who believes, as I do, that all the corporeal and mental organs (excepting those which are neither advantageous nor disadvantageous to the possessor) of all beings have been developed through natural selection, or the survival of the fittest, together with use or habit, will admit that these organs have been formed so that their possessors may compete successfully with other beings, and thus increase in number. Now an animal may be led to pursue that course of action which is most beneficial to the species by suffering, such as pain, hunger, thirst, and fear; or by pleasure, as in eating and drinking, and in the propagation of the species, etc.; or by both means combined, as in the search for food. But pain or suffering of any kind, if long continued, causes depression and lessens the power of action, yet is well adapted to make a creature guard itself against any great or sudden evil. Pleasurable sensations, on the other hand, may be long continued without any depressing effect; on the contrary, they stimulate the whole system to increased action. Hence it has come to pass that most or all sentient beings have been developed in such a manner, through natural selection, that pleasurable sensations serve as their habitual guides. We see this in the pleasure from exertion, even occasionally from great exertion of the body or mind—in the pleasure of our daily meals, and especially in the pleasure derived from sociability, and from loving our families. The sum of such pleasures as these, which are habitual or frequently recurrent, give, as I can hardly doubt, to most sentient beings an excess of happiness over misery, although many occasionally suffer much. Such suffering is quite compatible with the belief in Natural Selection, which is not perfect in its action, but tends only to render each species as successful as possible in the battle for life with other species, in wonderfully complex and changing circumstances.

That there is much suffering in the world no one disputes. Some have attempted to explain this with reference to man by imagining that it serves for his moral improvement. But the number of men in the world is as nothing compared with that of all other sentient beings, and they often suffer greatly without any moral improvement. This very old argument from the existence of suffering against the existence of an intelligent First Cause seems to me a strong one; whereas, as just remarked, the presence of much suffering agrees well with the view that all organic beings have been developed through variation and natural selection.

At the present day the most usual argument for the existence of an intelligent God is drawn from the deep inward conviction and feelings which are experienced by most persons.

Formerly I was led by feelings such as those just referred to (although I do not think that the religious sentiment was ever strongly developed in me), to the firm conviction of the existence of God, and of the immortality of the soul. In my Journal I wrote that whilst standing in the midst of the grandeur of a Brazilian forest, "it is not possible to give an adequate idea of the higher feelings of wonder, admiration, and devotion, which fill and elevate the mind." I well remember my conviction that there is more in man than the mere breath of his body. But now the grandest scenes would not cause any such convictions and feelings to rise in my mind. It may be truly said that I am like a man who has become color-blind, and the universal belief by men of the existence of redness makes my present loss of percep-

A New, Revolutionary World View

tion of not the least value as evidence. This argument would be a valid one if all men of all races had the same inward conviction of the existence of one God; but we know that this is very far from being the case. Therefore I cannot see that such inward convictions and feelings are of any weight as evidence of what really exists. The state of mind which grand scenes formerly excited in me, and which was intimately connected with a belief in God, did not essentially differ from that which is often called the sense of sublimity; and however difficult it may be to explain the genesis of this sense, it can hardly be advanced as an argument for the existence of God, any more than the powerful though vague and similar feelings excited by music.

With respect to immortality, nothing shows me [so clearly] how strong and almost instinctive a belief it is, as the consideration of the view now held by most physicists, namely, that the sun with all the planets will in time grow too cold for life, unless indeed some great body dashes into the sun, and thus gives it fresh life. Believing as I do that man in the distant future will be a far more perfect creature than he now is, it is an intolerable thought that he and all other sentient beings are doomed to complete annihilation after such long-continued slow progress. To those who fully admit the immortality of the human soul, the destruction of our world will not appear so dreadful.

Another source of conviction in the existence of God, connected with the reason, and not with the feelings, impresses me as having much more weight. This follows from the extreme difficulty or rather impossibility of conceiving this immense and wonderful universe, including man with his capacity of looking far backwards and far into futurity, as the result of blind chance or necessity. When thus reflecting I feel compelled to look to a First Cause having an intelligent mind in some degree analogous to that of man; and I deserve to be called a Theist. This conclusion was strong in my mind about the time, as far as I can remember, when I wrote the *Origin of Species;* and it is since that time that it has very gradually, with many fluctuations, become weaker. But then arises the doubt, can the mind of man, which has, as I fully believe, been developed from a mind as low as that possessed by the lowest animals, be trusted when it draws such grand conclusions?

I cannot pretend to throw the least light on such abstruse problems. The mystery of the beginning of all things is insoluble by us; and I for one must be content to remain an Agnostic.

NATURE AS BLUNDERING, LOW, AND HORRIBLY CRUEL

Nothing is easier than to admit in words the truth of the universal struggle for life, or more difficult—at least I have found it so—than constantly to bear this conclusion in mind. Yet unless it be thoroughly engrained in the mind, I am convinced that the whole economy of nature, with every fact on distribution, rarity, abundance, extinction, and variation, will be dimly seen or quite misunderstood. We behold the face of nature bright with gladness, we often see superabundance of food; we do not see, or we forget, that the birds which are idly singing round us mostly live on insects or seeds, and are thus constantly destroying life; or we forget how largely these songsters, or their eggs, or their nestlings, are destroyed by birds and beasts of prey; we do not always bear in mind, that though food may be now superabundant, it is not so at all seasons of each recurring year.

I should premise that I use the term Struggle for Existence in a large and metaphorical sense, including dependence of one being on another, and including (which is more important) not only the life of the individual, but success in leaving progeny. Two canine animals in a time of dearth, may be truly said to struggle with each other which shall get food and live. But a plant on the edge of a desert is said to struggle for life against the drought, though more properly it should be said to be dependent on the moisture. A plant which annually produces a thousand seeds, of which on an average only one comes to maturity, may be more truly said to struggle with the plants of the same and other kinds which already clothe the ground. The mistletoe is dependent on the apple and a few other trees, but can only in a far-fetched sense be said to struggle with these trees, for if too many of these parasites grow on the same tree, it will languish and die. But several seedling mistletoes, growing close together on the same branch, may more truly be said to struggle

Abridged from Charles Darwin, *On the Origin of Species* (1st ed.; London: John Murray, 1859).

A New, Revolutionary World View

with each other. As the mistletoe is disseminated by birds, its existence depends on birds; and it may metaphorically be said to struggle with other fruit-bearing plants, in order to tempt birds to devour and thus disseminate its seeds rather than those of other plants. In these several senses, which pass into each other, I use for convenience sake the general term of struggle for existence.

There is no exception to the rule that every organic being naturally increases at so high a rate, that if not destroyed, the earth would soon be covered by the progeny of a single pair. Even slow-breeding man has doubled in twenty-five years, and at this rate, in a few thousand years, there would literally not be standing room for his progeny. Linnaeus has calculated that if an annual plant produced only two seeds—and there is no plant so unproductive as this—and their seedlings next year produced two, and so on, then in twenty years there would be a million plants. The elephant is reckoned to be the slowest breeder of all known animals, and I have taken some pains to estimate its probable minimum rate of natural increase: it will be under the mark to assume that it breeds when thirty years old, and goes on breeding till ninety years old, bringing forth three pair of young in this interval; if this be so, at the end of the fifth century there would be alive fifteen million elephants, descended from the first pair.

But we have better evidence on this subject than mere theoretical calculations, namely, the numerous recorded cases of the astonishingly rapid increase of various animals in a state of nature, when circumstances have been favorable to them during two or three following seasons. Still more striking is the evidence from our domestic animals of many kinds which have run wild in several parts of the world: if the statements of the rate of increase of slow-breeding cattle and horses in South America, and latterly in Australia, had not been well authenticated, they would have been quite incredible. So it is with plants: cases could be given of introduced plants which have become common throughout whole islands in a period of less than ten years.

In a state of nature almost every plant produces seed, and amongst animals there are very few which do not annually pair. Hence we may confidently assert, that all plants and animals are tending to increase at a geometrical ratio, that all would most rapidly stock every station in which they could anyhow exist, and that the geometrical tendency to increase must be checked by destruction at some period of life. Our familiarity with the larger domestic animals tends, I think, to mislead us: we see no great destruction falling on them, and we forget that thousands are annually slaughtered for food, and that in a state of nature an equal number would have somehow to be disposed of.

A large number of eggs is of some importance to those species, which depend on a rapidly fluctuating amount of food, for it allows them rapidly to increase in number. But the real importance of a large number of eggs or seeds is to make up for much destruction at some period of life; and this period in the great majority of cases is an early one. If an animal can in any way protect its own eggs or young, a small number may be produced, and yet the average stock be fully kept up; but if many eggs or young are destroyed, many must be produced, or the species will become extinct. It would suffice to keep up the full number of a tree, which lived on an average for a thousand years, if a single seed were produced once in a thousand years, supposing that this seed were never destroyed, and could be ensured to germinate in a fitting place. So that in all cases, the average number of any animal depends only indirectly on the number of its eggs or seeds.

In looking at Nature, it is most necessary to keep the foregoing considerations always in mind—never to forget that every single organic being around us may be said to be striving to the utmost to increase in numbers; that each lives by a struggle at some period of its life; that heavy destruction inevitably falls either on the young or old, during each generation or at recurrent intervals. Lighten any check, mitigate the destruction ever so little, and the number of the species will almost instantaneously increase to any amount. The face of Nature may be compared to a yielding surface, with ten thousand sharp wedges packed close together and driven inwards by incessant blows, sometimes one wedge being struck, and then another with greater force.

The amount of food for each species of course gives the extreme limit to which each can increase; but very frequently it is not the obtaining food, but the serving as prey to other animals, which determines the average numbers of a species. Thus, there seems to be little doubt that the stock of partridges, grouse, and hares, on any large estate, depends chiefly on the destruction of vermin. If not one head of game were shot during the next twenty years in England,

and, at the same time, if no vermin were destroyed, there would, in all probability, be less game than at present, although hundreds of thousands of game animals are now annually killed. On the other hand, in some cases, as with the elephant and rhinoceros, none are destroyed by beasts of prey: even the tiger in India most rarely dares to attack a young elephant protected by its dam.

Climate plays an important part in determining the average numbers of a species, and periodical seasons of extreme cold or drought, I believe to be the most effective of all checks. I estimated that the winter of 1854–55 destroyed four-fifths of the birds in my own grounds; and this is a tremendous destruction, when we remember that 10 per cent is an extraordinarily severe mortality from epidemics with man. The action of climate seems at first sight to be quite independent of the struggle for existence; but insofar as climate chiefly acts in reducing food, it brings on the most severe struggle between the individuals, whether of the same or of distinct species, which subsist on the same kind of food. Even when climate, for instance extreme cold, acts directly, it will be the least vigorous, or those which have got least food through the advancing winter, which will suffer most.

Many cases are on record showing how complex and unexpected are the checks and relations between organic beings, which have to struggle together in the same country. I will give only a single instance, which, though a simple one, has interested me. In Staffordshire, on the estate of a relation where I had ample means of investigation, there was a large and extremely barren heath, which had never been touched by the hand of man; but several hundred acres of exactly the same nature had been enclosed twenty-five years previously and planted with Scotch fir. The change in the native vegetation of the planted part of the heath was most remarkable, more than is generally seen in passing from one quite different soil to another: not only the proportional numbers of the heath plants were wholly changed, but twelve species of plants (not counting grasses and carices) flourished in the plantations, which could not be found on the heath. The effect on the insects must have been still greater, for six insectivorous birds were very common in the plantations, which were not to be seen on the heath; and the heath was frequented by two or three distinct insectivorous birds. Here we see how potent has been the effect of the introduction of a single tree, nothing whatever else having been done, with the exception that the land had been enclosed, so that cattle could not enter. But how important an element enclosure is, I plainly saw near Farnham, in Surrey. Here there are extensive heaths, with a few clumps of old Scotch firs on the distant hilltops: within the last ten years large spaces have been enclosed, and self-sown firs are now springing up in multitudes, so close together that all cannot live. When I ascertained that these young trees had not been sown or planted, I was so much surprised at their numbers that I went to several points of view, whence I could examine hundreds of acres of the unenclosed heath, and literally I could not see a single Scotch fir, except the old planted clumps. But on looking closely between the stems of the heath, I found a multitude of seedlings and little trees, which had been perpetually browsed down by the cattle. In one square yard, at a point some hundred yards distant from one of the old clumps, I counted thirty-two little trees; and one of them, judging from the rings of growth, had during twenty-six years tried to raise its head above the stems of the heath, and had failed. No wonder that, as soon as the land was enclosed, it became thickly clothed with vigorously growing young firs. Yet the heath was so extremely barren and so extensive that no one would ever have imagined that cattle would have so closely and effectually searched it for food.

I am tempted to give one more instance showing how plants and animals, most remote in the scale of nature, are bound together by a web of complex relations. . . . From experiments which I have tried, I have found that the visits of bees, if not indispensable, are at least highly beneficial to the fertilization of our clovers; but humble-bees alone visit the common red clover (*Trifolium pratense*), as other bees cannot reach the nectar. Hence I have very little doubt, that if the whole genus of humble-bees became extinct or very rare in England, the heartsease and red clover would become very rare, or wholly disappear. The number of humble-bees in any district depends in a great degree on the number of fieldmice, which destroy their combs and nests; and Mr. H. Newman, who has long attended to the habits of humble-bees, believes that "more than two-thirds of them are thus destroyed all over England." Now the number of mice is largely dependent, as every one knows, on the number of cats; and Mr. Newman says, "Near villages and small towns I have found the

A New, Revolutionary World View

nests of humble-bees more numerous than elsewhere, which I attribute to the number of cats that destroy the mice." Hence it is quite creditable that the presence of a feline animal in large numbers in a district might determine, through the intervention first of mice and then of bees, the frequency of certain flowers in that district!

In the case of every species, many different checks, acting at different periods of life, and during different seasons or years, probably come into play; some one check or some few being generally the most potent, but all concurring in determining the average number or even the existence of the species. In some cases it can be shown that widely different checks act on the same species in different districts. When we look at the plants and bushes clothing an entangled bank, we are tempted to attribute their proportional numbers and kinds to what we call chance. But how false a view is this! Everyone has heard that when an American forest is cut down, a very different vegetation springs up; but it has been observed that the trees now growing on the ancient Indian mounds, in the southern United States, display the same beautiful diversity and proportion of kinds as in the surrounding virgin forests. What a struggle between the several kinds of trees must here have gone on during long centuries, each annually scattering its seeds by the thousand; what war between insect and insect—between insects, snails, and other animals with birds and beasts of prey—all striving to increase, and all feeding on each other or on the trees or their seeds and seedlings, or on the plants which first clothed the ground and thus checked the growth of the trees! Throw up a handful of feathers, and all must fall to the ground according to definite laws; but how simple is this problem compared to the action and reaction of the innumerable plants and animals which have determined, in the course of centuries, the proportional numbers and kinds of trees now growing on the old Indian ruins!

The dependency of one organic being on another, as of a parasite on its prey, lies generally between beings remote in the scale of nature. This is often the case with those which may strictly be said to struggle with each other for existence, as in the case of locusts and grass-feeding quadrupeds. But the struggle almost invariably will be most severe between the individuals of the same species, for they frequent the same districts, require the same food, and are exposed to the same dangers. In the case of varieties of the same species, the struggle will generally be almost equally severe, and we sometimes see the contest soon decided: for instance, if several varieties of wheat be sown together, and the mixed seeds be resown, some of the varieties which best suit the soil or climate, or are naturally the most fertile, will beat the others and so yield more seed, and will consequently in a few years quite supplant the other varieties. To keep up a mixed stock of even such extremely close varieties as the variously colored sweet-peas, they must be each year harvested separately, and the seed then mixed in due proportion, otherwise the weaker kinds will steadily decrease in numbers and disappear. So again with the varieties of sheep: it has been asserted that certain mountain varieties will starve out other mountain varieties, so that they cannot be kept together.

A corollary of the highest importance may be deduced from the foregoing remarks, namely, that the structure of every organic being is related, in the most essential yet often hidden manner, to that of all other organic beings, with which it comes into competition for food or residence, or from which it has to escape, or on which it preys. This is obvious in the structure of the teeth and talons of the tiger; and in that of the legs and claws of the parasite which clings to the hair on the tiger's body. But in the beautifully plumed seed of the dandelion, and in the flattened and fringed legs of the water-beetle, the relation seems at first confined to the elements of air and water. Yet the advantage of plumed seeds no doubt stands in the closest relation to the land being already thickly clothed by other plants; so that the seeds may be widely distributed and fall on unoccupied ground. In the water-beetle, the structure of its legs, so well adapted for diving, allows it to compete with other aquatic insects, to hunt for its own prey, and to escape serving as prey to other animals.

The store of nutriment laid up within the seeds of many plants seems at first sight to have no sort of relation to other plants. But from the strong growth of young plants produced from such seeds (as peas and beans), when sown in the midst of long grass, I suspect that the chief use of the nutriment in the seed is to favor the growth of the young seedling, whilst struggling with other plants growing vigorously all around.

A New, Revolutionary World View

Look at a plant in the midst of its range, why does it not double or quadruple its numbers? We know that it can perfectly well withstand a little more heat or cold, dampness or dryness, for elsewhere it ranges into slightly hotter, or colder, damper or drier districts. In this case we can clearly see that if we wished in imagination to give the plant the power of increasing in number, we should have to give it some advantage over its competitors, or over the animals which preyed on it.

It is good thus to try in our imagination to give any form some advantage over another. Probably in no single instance should we know what to do, so as to succeed. It will convince us of our ignorance on the mutual relations of all organic beings; a conviction as necessary, as it seems to be difficult to acquire. All that we can do, is to keep steadily in mind that each organic being is striving to increase at a geometrical ratio; that each at some period of its life, during some season of the year, during each generation or at intervals, has to struggle for life, and to suffer great destruction. When we reflect on this struggle, we may console ourselves with the full belief that the war of nature is not incessant, that no fear is felt, that death is generally prompt, and that the vigorous, the healthy, and the happy survive and multiply.

* * *

It is interesting to contemplate an entangled bank, clothed with many plants of many kinds, with birds singing on the bushes, with various insects flitting about, and with worms crawling through the damp earth, and to reflect that these elaborately constructed forms, so different from each other, and dependent on each other in so complex a manner, have all been produced by laws acting around us. These laws, taken in the largest sense, being Growth with Reproduction; Inheritance which is almost implied by reproduction; Variability from the indirect and direct action of the external conditions of life, and from use and disuse; a Ratio of Increase so high as to lead to a Struggle for Life, and as a consequence to Natural Selection, entailing Divergence of Character and the Extinction of less-improved forms. Thus, from the war of nature, from famine and death, the most exalted object which we are capable of conceiving, namely, the production of the higher animals, directly follows. There is a grandeur in this view of life, with its several powers, having been originally breathed into a few forms or into one; and that, whilst this planet has gone cycling on according to the fixed law of gravity, from so simple a beginning endless forms most beautiful and most wonderful have been, and are being, evolved.

Samuel Wilberforce and Adam Sedgwick
TRADITIONAL RELIGION AND SCIENCE OPPOSE EVOLUTION

The most momentous public clash between a Darwinian and a traditionalist occurred at Oxford before the British Association of Science, shortly after the publication of The Origin of Species. *Bishop Samuel Wilberforce (1805–1872), advised by one of Darwin's scientific opponents, Richard Owen, depicted Darwin's research as a "rotten fabric of guess and speculation" that betrayed inductive science and debased man, God, and nature. Wilberforce's aspersions were matched by the vigorous reply of T. H. Huxley, who said that the bishop should stay in "his own sphere of activity" instead of indulging in "eloquent digressions, and skilled appeals to religious prejudice."*

The first selection below, entitled "Bishop Wilberforce's Review," is taken from Wilberforce's anonymous, lengthy analysis of the Origin. *This review was published after the Oxford exchange and indicates how traditional religion and science were used both to criticize and decry Darwinism. The second reading, "Sedgwick's Objections," shows how the once-distinguished, progressive geologist Adam Sedgwick (1785–1873) lost his composure and a good measure of his famous wit upon reading* The Origin of Species. *Like Wilberforce (and behind him, the lionized anatomist and paleontologist, Richard Owen), Sedgwick accused Darwin of forsaking the inductive method. Sedgwick then sought to prove that geological evidence supported the ideas of creation, fixity of species, and geological development via catastrophic periods of change, not Darwin's theory of "transmutation." Yet beyond the methods and data of science, Sedgwick was most disturbed by the way Darwin seemed to separate science from morality and religion, and to degrade humankind.*

BISHOP WILBERFORCE'S REVIEW

Any contribution to our Natural History literature from the pen of Mr. C. Darwin is certain to command attention. His scientific attainments, his insight and carefulness as an observer, blended with no scanty measure of imaginative sagacity, and his clear and lively style, make all his writings unusually attractive. His present volume on the "Origin of Species" is the result of many years of observation,

---

Abridged from Samuel Wilberforce, "On the Origin of Species," *Quarterly Review* 108 (1860).

thought, and speculation; and is manifestly regarded by him as the "opus" upon which his future fame is to rest.

The essay is full of Mr. Darwin's characteristic excellences. It is a most readable book; full of facts in natural history, old and new, of his collecting and of his observing; and all of these are told in his own perspicuous language, and all thrown into picturesque combinations, and all sparkle with the colors of fancy and the lights of imagination. It assumes, too, the grave proportions of a sustained argument upon a matter of the deepest interest, not to naturalists only, or even to men of science exclusively, but to everyone who is interested in the history of man and of the relations of nature around him to the history and plan of creation.

With Mr. Darwin's "argument" we may say in the outset that we shall have much and grave fault to find. But this does not make us the less disposed to admire the singular excellences of his work.

We feel as we walk abroad with Mr. Darwin very much as the favored object of the attention of the dervish must have felt when he had rubbed the ointment around his eye, and had it opened to see all the jewels, and diamonds, and emeralds, and topazes, and rubies, which were sparkling unregarded beneath the earth, hidden as yet from all eyes save those which the dervish had enlightened. But here we are bound to say our pleasure terminates; for, when we turn with Mr. Darwin to his "argument," we are almost immediately at variance with him. It is as an "argument" that the essay is put forward; as an argument we will test it.

The conclusion . . . to which Mr. Darwin would bring us is, that all the various forms of vegetable and animal life with which the globe is now peopled, or of which we find the remains preserved in a fossil state in the great Earth-Museum around us, which the science of geology unlocks for our instruction, have come down by natural succession of descent from father to son—"animals from at most four or five progenitors, and plants from an equal or less number" (p. 484), as Mr. Darwin at first somewhat diffidently suggests. . . . This is the theory which really pervades the whole volume. Man, beast, creeping thing, and plant of the earth, are all the lineal and direct descendants of some one individual *ens,* whose various progeny have been simply modified by the action of natural and ascertainable conditions into the multiform aspect of life which we see around us. This is undoubtedly at first sight a somewhat startling conclusion to arrive at. To find that mosses, grasses, turnips, oaks, worms, and flies, mites and elephants, infusoria and whales, tadpoles of today and venerable saurians, truffles and men, are all equally the lineal descendants of the same aboriginal common ancestor, perhaps of the nucleated cell of some primeval fungus, which alone possessed the distinguishing honor of being the "one primordial form into which life was first breathed by the Creator"—this, to say the least of it, is no common discovery—no very expected conclusion. But we are too loyal pupils of inductive philosophy to start back from any conclusion by reason of its strangeness. Newton's patient philosophy taught him to find in the falling apple the law which governs the silent movements of the stars in their courses; and if Mr. Darwin can with the same correctness of reasoning demonstrate to us our fungular descent, we shall dismiss our pride, and avow, with the characteristic humility of philosophy, our unsuspected cousinship with the mushrooms—

"Claim kindred there, and have our claim allowed,"

—only we shall ask leave to scrutinize carefully every step of the argument which has such an ending, and demur if at any point of it we are invited to substitute unlimited hypothesis for patient observation, or the spasmodic fluttering flight of fancy for the severe conclusions to which logical accuracy of reasoning has led the way.

Now, the main propositions by which Mr. Darwin's conclusion is attained are these:

1. That observed and admitted variations spring up in the course of descents from a common progenitor.
2. That many of these variations tend to an improvement upon the parent stock.
3. That, by a continued selection of these improved specimens as the progenitors of future stock, its powers may be unlimitedly increased.
4. And, lastly, that there is in nature a power continually and universally working out this selection, and so fixing and augmenting these improvements.

Mr. Darwin's whole theory rests upon the truth of these propositions, and crumbles utterly away if only one of them fail him. These therefore we must closely scrutinize. We will begin with the last in our series, both because we think it the newest and the most ingenious part of Mr. Darwin's whole argument, and also because, whilst we absolutely deny the mode in which he seeks to apply the existence of the power to help him in his argument, yet we think that he throws great and very interesting light upon the fact that such a self-acting power does actively and continuously work in all creation around us.

Mr. Darwin finds then the disseminating and improving power, which he needs to account for the development of new forms in nature, in the principle of "Natural Selection," which is evolved in the strife for room to live and flourish which is evermore maintained between themselves by all living things. One of the most interesting parts of Mr. Darwin's volume is that in which he establishes this law of natural selection; we say establishes, because—repeating that we differ from him totally in the limits which he would assign to its action—we have no doubt of the existence or of the importance of the law itself.

That such a struggle for life then actually exists, and that it tends continually to lead the strong to exterminate the weak, we readily admit; and in this law we see a merciful provision against the deterioration, in a world apt to deteriorate, of the works of the Creator's hands. Thus it is that the bloody strifes of the males of all wild animals tend to maintain the vigor and full development of their race; because, through this machinery of appetite and passion, the most vigorous individuals become the progenitors of the next generation of the tribe. And this law, which thus maintains through the struggle of individuals the high type of the family, tends continually, through a similar struggle of species, to lead the stronger species to supplant the weaker.

But before we can go a step further, and argue from its operation in favor of a perpetual improvement in natural types, we must be shown first that this law of competition has in nature to deal with such favorable variations in the individuals of any species, as truly to exalt those individuals above the highest type of perfection to which their least imperfect predecessors attained—above, that is to say, the normal level of the species . . . and then, next, we must be shown that there is actively at work in nature, coordinate with the law of competition and with the existence of such favorable variations, a power of accumulating such favorable variation through successive descents. Failing the establishment of either of these last two propositions, Mr. Darwin's whole theory falls to pieces. He has accordingly labored with all his strength to establish these, and into that attempt we must now follow him.

Mr. Darwin begins by endeavoring to prove that such variations are produced under the selecting power of man amongst domestic animals. . . . He writes a delightful chapter upon pigeons. Runts and fantails, short-faced tumblers and long-faced tumblers, long-beaked carriers and pouters, black barbs, jacobins, and turbits, coo and tumble, inflate their esophagi, and pout and spread out their tails before us. We learn that "pigeons have been watched and tended with the utmost care, and loved by many people." They have been domesticated for thousands of years in several quarters of the world.

Now all this is very pleasant writing, especially for pigeon-fanciers; but what step do we really gain in it at all towards establishing the alleged fact that variations are but species in the act of formation, or in establishing Mr. Darwin's position that a well-marked variety may be called an incipient species? We affirm positively that no single *fact* tending even in that direction is brought forward. On the contrary, every one points distinctly towards the opposite conclusion; for with all the change wrought in appearance, with all the apparent variation in manners, there is not the faintest beginning of any such change in what that great comparative anatomist, Professor Owen, calls "the characteristics of the skeleton or other parts of the frame upon which specific differences are founded."

Nor must we pass over unnoticed the transference of the argument from the domesticated to the untamed animals. Assuming that man as the selector can do much in a limited time, Mr. Darwin argues that Nature, a more powerful, a more continuous power, working over vastly extended ranges of time, can do more. But why should Nature, so uniform and persistent in all her operations, tend in this instance to change? Why should she become a selector of varieties? Because, most ingeniously argues Mr. Darwin, in the struggle for life, *if* any variety favorable to the individual were developed, that individual would have a better chance in the battle of life, would assert more proudly his own place, and, handing on his peculiarity to his descendants, would become the progenitor of an improved race; and so a variety would have grown into a species.

We think it difficult to find a theory fuller of assumptions; and of assumptions not grounded upon alleged facts in nature, but which are absolutely opposed to all the facts we have been able to observe.

The applied argument then, from variation under domestication, fails utterly. But further, what does observation say as to the occurrence of a single instance of such favorable variation? Men have now for thousands of years been conversant as hunters and other rough naturalists with animals of every class. Has any one such instance ever been discovered? We fearlessly assert not one.

Here then again, when subjected to the stern Baconian law of the observation of facts, the theory breaks down utterly; for no natural variations from the specific type favorable to the individual from which nature is to select can anywhere be found.

We come then to these conclusions. All the facts presented to us in the natural world tend to show that none of the variations produced in the fixed forms of animal life, when seen in its most plastic condition under domestication, give any promise of a true transmutation of species; first, from the difficulty of accumulating and fixing variations within the same species; secondly, from the fact that these variations, though most serviceable for man, have no tendency to improve the individual beyond the standard of his own specific type, and so to afford matter, even if they were infinitely produced, for the supposed power of natural selection on which to work; whilst all variations from the mixture of species are barred by the inexorable law of hybrid sterility. Further, the embalmed records of 3000 years show that there has been no beginning of transmutation in the species of our most familiar domesticated animals; and beyond this, that in the countless tribes of animal life around us, down to its lowest and most variable species, no one has ever discovered a single instance of such transmutation being now in prospect; no new organ has ever been known to be developed—no new natural instinct to be formed—whilst, finally, in the vast museum of departed animal life which the strata of the earth imbed for our examination, whilst they contain far too complete a representation of the past to be set aside as a mere imperfect record, yet afford no one instance of any such change as having ever been in progress, or give us anywhere the missing links of the assumed chain, or the remains which would enable now existing variations, by gradual approximations, to shade off into unity.

On what then is the new theory based? We say it with unfeigned regret, in dealing with such a man as Mr. Darwin, on the merest hypothesis, supported by the most unbounded assumptions.

Sometimes Mr. Darwin seems for a moment to recoil himself from this extravagant liberty of speculation, as when he says, concerning the eye,

> To suppose that the eye, with its inimitable contrivances for adjusting the focus to different distances, for admitting different amounts of light, and for the correction of spherical and chromatic aberration, could have been formed by natural selection, seems, I freely confess, absurd in the highest possible degree.—p. 186.

But he soon returns to his new wantonness of conjecture, and, without the shadow of a fact, contents himself with saying that

> he suspects that any sensitive nerve may be rendered sensitive to light, and likewise to those coarser vibrations of the air which produce sound. —p. 187.

And in the following passage he carries this extravagance to the highest pitch, requiring a license for advancing as true any theory which cannot be demonstrated to be actually impossible:

> If it could be demonstrated that any complex organ existed, which could not possibly have been formed by numerous, successive, slight modifications, my theory would absolutely break down. But I can find no such case.—p. 189. . . .

In the name of all true philosophy we protest equally against such a mode of dealing with nature, as utterly dishonorable to all natural science, as reducing it from its present lofty level as one of the noblest trainers of man's intellect and instructors of his mind, to being a mere idle play of the fancy, without the basis of fact or the discipline of observation. In the "Arabian Nights" we are not offended as at an impossibility when Amina sprinkles her husband with water and transforms him into a dog, but we cannot open the august doors of the venerable temple of scientific truth to the genii and magicians of romance.

Our readers will not have failed to notice that we have objected to the views with which we have been dealing solely on scientific grounds. We have done so from our fixed conviction that it is thus that the truth or falsehood of such arguments should be tried. We have no sympathy with those who object to any facts or alleged facts in nature, or to any inference logically deduced from them, because they believe them to contradict what it appears to them is taught by Revelation. We think that all such objections savour of a timidity which is really inconsistent with a firm and well-instructed faith. . . . To oppose facts in the natural world because they seem to oppose Revelation, or to humor them so as to compel them to speak its voice, is, he knows, but another form of the ever-ready feebleminded dishonesty of lying for God, and trying by fraud or falsehood to do the work of the God of truth. It is with another and a nobler spirit that the true believer walks amongst the works of nature. The words graven on the everlasting rocks are the words of God, and they are graven by His hand. No more can they contradict His Word written in His book, than could the words of the old covenant graven by His hand on the stony tables contradict the writings of His hand in the volume of the new dispensation. There may be to man difficulty in reconciling all the utterances of the two voices. But what of that? He has learned already that here he knows only in part, and that the day of reconciling all apparent contradictions between what must agree is nigh at hand. He rests his mind in perfect quietness on this assurance, and rejoices in the gift of light without a misgiving as to what it may discover:

> "A man of deep thought and great practical wisdom," says Sedgwick, "one whose piety and benevolence have for many years been shining before the world, and of whose sincerity no scoffer (of whatever school) will dare to start a doubt, recorded his opinion in the great assembly of the men of science who during the past year were gathered from every corner of the Empire within the walls of this University, 'that Christianity had everything to hope and nothing to fear from the advancement of philosophy.'"

We cannot, therefore, consent to test the truth of natural science by the Word of Revelation. But this does not make it the less important to point out on scientific grounds scientific errors, when those errors tend to limit God's glory in creation, or to gainsay the revealed rela-

Traditional Religion and Science Oppose Evolution

tions of that creation to Himself. To both these classes of error, though, we doubt not, quite unintentionally on his part, we think that Mr. Darwin's speculations directly tend.

Mr. Darwin writes as a Christian, and we doubt not that he is one. We do not for a moment believe him to be one of those who retain in some corner of their hearts a secret unbelief which they dare not vent; and we therefore pray him to consider well the grounds on which we brand his speculations with the charge of such a tendency. First, then, he not obscurely declares that he applies his scheme of the action of the principle of natural selection to MAN himself, as well as to the animals around him. Now, we must say at once, and openly, that such a notion is absolutely incompatible not only with single expressions in the word of God on that subject of natural science with which it is not immediately concerned, but, which in our judgment is of far more importance, with the whole representation of that moral and spiritual condition of man which is its proper subject matter. Man's derived supremacy over the earth; man's power of articulate speech; man's gift of reason; man's free-will and responsibility; man's fall and man's redemption; the incarnation of the Eternal Son; the indwelling of the Eternal Spirit—all are equally and utterly irreconcilable with the degrading notion of the brute origin of him who was created in the image of God, and redeemed by the Eternal Son assuming to himself his nature. Equally inconsistent, too, not with any passing expressions, but with the whole scheme of God's dealings with man as recorded in His word, is Mr. Darwin's daring notion of man's further development into some unknown extent of powers, and shape, and size, through natural selection acting through that long vista of ages which he casts mistily over the earth upon the most favored individuals of his species. We care not in these pages to push the argument further. We have done enough for our purpose in thus succinctly intimating its course.

Nor can we doubt, secondly, that this view, which thus contradicts the revealed relation of creation to its Creator, is equally inconsistent with the fulness of His glory. It is, in truth, an ingenious theory for diffusing throughout creation the working and so the personality of the Creator. And thus, however unconsciously to him who holds them, such views really tend inevitably to banish from the mind most of the peculiar attributes of the Almighty.

How, asks Mr. Darwin, can we possibly account for the manifest plan, order, and arrangement which pervade creation, except we allow to it this self-developing power through modified descent?

How can we account for all this? By the simplest and yet the most comprehensive answer. By declaring the stupendous fact that all creation is the transcript in matter of ideas eternally existing in the mind of the Most High—that order in the utmost perfectness of its relation pervades His works, because it exists as in its center and highest fountainhead in Him the Lord of all.

We think that the real temper of this whole speculation as to nature itself may be read in these few lines. It is a dishonouring view of nature.

That reverence for the work of God's hands with which a true belief in the All-wise Worker fills the believer's heart is at the root of all great physical discovery; it is the basis of philosophy. He who would see the venerable features of Nature must not seek with the rudeness of a licensed roysterer violently to unmask her countenance; but must wait as a learner for her willing unveiling. . . . But we can give . . . a simpler solution still for the presence of these strange forms of imperfection and suffering amongst the works of God.

We can tell him of the strong shudder which ran through all this world when its head and ruler fell. . . . Having shown that man is not and cannot be an improved ape, Professor Owen adds:

> . . . Man is the sole species of his genus, the sole representative of his order and subclass. Thus I trust has been furnished the confutation of the notion of a transformation of the ape into the man, which appears from a favorite old author to have been entertained by some in his day:
> "And of a truth, vile epicurism and sensuality will make the soul of man so degenerate and blind, that he will not only be content to slide into brutish immorality, but please himself in this very opinion that he is a real brute already, an ape, satyr, or baboon; and that the best of men are no better, saving that civilizing of them and industrious education has made them appear in a more refined shape, and long inculcated precepts have been mistaken for connate principles of honesty and natural knowledge; otherwise there be no indispensable grounds of religion and virtue but what has happened to be taken up by over-ruling custom, which things, I dare say, are as easily confutable as any conclusion in mathematics is demonstrable. But as many as are thus sottish, let them enjoy their own wildness and ignorance; it is sufficient for a good man that he is conscious unto himself that he is more nobly descended, better bred and born, and more skillfully taught by the purged faculties of his own mind."—Owen's Classification of Mammals, p. 103.

And he draws these truly philosophical views to this noble conclusion.

> Such are the dominating powers with which we, and we alone, are gifted! I say gifted, for the surpassing organization was no work of ours. It is He that hath made us, not we ourselves. This frame is a temporary trust, for the use of which we are responsible to the Maker. Oh! you who possess it in all the supple vigour of lusty youth, think well what it is that He has committed to your keeping. Waste not its energies; dull them not by sloth; spoil them not by pleasures!
>
> The supreme work of creation has been accomplished that you might possess a body—the sole erect—of all animal bodies the most free—and for what? For the service of the soul.

SEDGWICK'S OBJECTIONS

I must in the first place observe that Darwin's theory is not *inductive*—not based on a series of acknowledged facts pointing to a *general conclusion*—not a proposition evolved out of the facts, logically, and of course including them. To use an old figure, I look on the theory as a vast pyramid resting on its apex, and that apex a mathematical point. The only facts he pretends to adduce, as true elements of proof, are the *varieties* produced by domestication, or the *human artifice* of cross-breeding. We all admit the varieties, and the very wide limits of variation, among domestic animals. How very unlike are poodles and greyhounds! Yet they are of one species. And how nearly alike are many animals—allowed to be of distinct species, on any acknowledged views of species. Hence there may have been very many blunders among naturalists, in the discrimination and enumeration of species. But this does not undermine the grand truth of nature, and the continuity of true species. Again, the varieties, built upon by Mr. Darwin, are varieties of domestication and human *design*. Such varieties could have no existence in the old world. Something may be done by cross-breeding; but mules are generally sterile, or the progeny (in some rare instances) passes into one of the original crossed forms. The Author of Nature will not permit His work to be spoiled by the wanton curiosity of Man. And in a state of nature (such as that of the old world before Man came upon it) wild animals of different species do not desire to cross and unite.

Abridged from Adam Sedgwick, "Objections to Mr. Darwin's Theory of the Origin of Species," *The Spectator* (April 7, 1860).

Species have been constant for thousands of years; and time (so far as I see my way) though multiplied by millions and billions would never change them, so long as the conditions remained constant. Change the conditions, and old species would disappear; and new species *might* have room to come in and flourish. But how, and by what causation? I say by *creation.* But, what do I mean by creation? I reply, the operation of a power quite beyond the powers of a pigeon-fancier, a cross-breeder, or hybridizer; a power I cannot imitate or comprehend, but in which I can believe, by a legitimate conclusion of sound reason drawn from the laws and harmonies of Nature. For I can see in all around me a design and purpose, and a mutual adaptation of parts, which I *can* comprehend—and which prove that there is exterior to, and above, the mere phenomena of Nature a great ... designing cause. Believing this, I have no difficulty in the repetition of new species during successive epochs in the history of the earth.

But Darwin would say I am introducing a *miracle* by the supposition. In one sense, I am; in another, I am not. The hypothesis does not suspend or interrupt an established law of Nature. It does suppose the introduction of a new phenomenon unaccounted for by the operation of any *known* law of Nature; and it appeals to a power above established laws, and yet acting in harmony and conformity with them.

The pretended physical philosophy of modern days strips Man of all his moral attributes, or holds them of no account in the estimate of his origin and place in the created world. A cold atheistical materialism is the tendency of the so-called material philosophy of the present day. Not that I believe that Darwin is an atheist; though I cannot but regard his materialism as atheistical; because it ignores all rational conception of a final cause. I think it untrue, because opposed to the obvious course of Nature, and the very opposite of inductive truth. I therefore think it intensely mischievous.

Let no one say that it is held together by a *cumulative* argument. Each series of facts is laced together by a series of assumptions, which are mere repetitions of the one false principle. You cannot make a good rope out of a string of air-bubbles.

I proceed now to notice the manner in which Darwin tries to fit his principles to the facts of geology.

I will take for granted that the known series of fossil-bearing rocks or deposits may be divided into the Paleozoic; the Mesozoic; the Ter-

Traditional Religion and Science Oppose Evolution 87

tiary or Neozoic; and the Modern—the Fens, Deltas, etc., etc., with the spoils of the actual flora and fauna of the world, and with wrecks of the works of Man.

To begin then, with the Paleozoic rocks. Surely we ought on the transmutation theory, to find near their base great deposits with *none but the lowest forms of organic life.* I know of no such deposits. Oken contends that life began with the infusorial forms. They are at any rate well fitted for fossil preservation, but we do not find them. Neither do we find beds exclusively of hard corals and other humble organisms, which ought, on the theory, to mark a period of vast duration while the primeval monads were working up into the higher types of life. Our evidence is, no doubt, very scanty; but let not our opponents dare to say that it makes *for them.* So far as it is positive, it seems to me point-blank *against them.* If *we* build upon imperfect evidence, *they* commence without any evidence whatsoever, and against the evidence of actual nature. As we ascend in the great stages of the Paleozoic series (through Cambrian, Silurian, Devonian, and Carboniferous rocks, we have in each a *characteristic* fauna), we have no wavering of species—we have the noblest cephalopods and brachiopods that ever existed, and they preserve their typical forms till they disappear. And a few of the types have endured, with specific modifications, through all succeeding ages of the earth. It is during these old periods that we have some of the noblest icthye forms that ever were created. The same may be said, I think, of the carboniferous flora. As a whole, indeed, it is lower than the living flora of our own period; but many of the old types were grander and of higher organization than the corresponding families of the living flora; and there is no wavering, no wanting of organic definition in the old types. We have some land reptiles (batrachians), in the higher Paleozoic periods, but not of a very low type; and the reptiles of the permian groups (at the very top of the Paleozoic rocks) are of a high type. If all this be true (and I think it is), it gives but a sturdy grist for the transmutation-mill, and may soon break its cogs.

We know the complicated organic phenomena of the Mesozoic (or Oolitic) period. It defies the transmutationist at every step. Oh! but the document, says Darwin, is a fragment. I will interpolate long periods to account for all the changes. I say, in reply, if you deny my conclusion grounded on positive evidence, I toss back your conclusions, derived from negative evidence—the inflated cushion on which you try to bolster up the defects of your hypothesis. The reptile fauna of the Mesozoic period is the grandest and highest that ever lived. How came these reptiles to die off, or to degenerate? And how came the Dinosaurs to disappear from the face of Nature, and leave no descendants like themselves, or of a corresponding nobility? By what process of *natural selection* did they disappear? Did they tire of the land, and become Whales, casting off their hindlegs? And, after they had lasted millions of years as whales, did they tire of the water, and leap out again as Pachyderms? I have heard of both hypotheses; and I cannot put them into words without seeming to use the terms of mockery. This I do affirm, that if the transmutation theory were proved true in the actual world, and we could hatch rats out of the eggs of geese, it would still be difficult to account for the successive forms of organic life in the old world. They appear to me to give the lie to the theory of transmutation at every turn of the pages of Dame Nature's old book.

The limits of this letter compel me to omit any long discussion of the Tertiary Mammals, of course including man at their head. On physical grounds, the transmutation theory is untrue, if we reason (as we ought to do) from the known to the unknown. To this rule, the Tertiary Mammals offer us no exception. Nor is there any proof, either ethnographical or physical, of the bestial origin of man.

And now for a few words upon Darwin's long *interpolated periods* of geological ages. He has an eternity of past time to draw upon; and I am willing to give him ample measure; only let him use it logically, and in some probable accordance with facts and phenomena.

Towards the end of the carboniferous period, there was a vast extinction of animal and vegetable life. We can, I think, account for this extinction mechanically. The old crust was broken up. The sea bottom underwent a great change. The old flora and fauna went out; and a new flora and fauna appeared, in the ground, now called permian, at the base of the new red sandstone, which overlies the carboniferous rocks. I take the fact as it *is,* and I have no difficulty. The time in which all this was brought about *may* have been very long, even upon a geological scale of time. But where do the *intervening* and connecting types exist, which are to mark the work of *natural selection*? We do not find them. Therefore, the step onwards gives no true resting-place to a baseless theory; and is, in fact, a stumbling-block in its way.

Before we rise through the new red sandstone, we find the muschel-kalk (wanting in England, though its place on the scale is well-known) with *an entirely new* fauna: where have we a proof of any enormous lapse of geological time to account for the change? We have no proof in the deposits themselves: the presumption they offer to our senses is of a contrary kind.

If we rise from the muschel-kalk to the lias, we find again a new fauna. All the anterior species are gone. Yet the passage through the upper members of the new red sandstone to the lias is by insensible gradations, and it is no easy matter to fix the physical line of their demarcation. I think it would be a very rash assertion to affirm that a great geological interval took place between the formation of the upper part of the new red sandstone and the lias. Physical evidence is against it. To support a baseless theory, Darwin would require a countless lapse of ages of which we have no commensurate physical monuments; and he is unable to supply any of the connecting organic links that ought to bind together the older fauna with that of the lias.

I cannot go on any further with these objections. But I will not conclude without expressing my deep aversion to the theory; because of its unflinching materialism;—because it has deserted the inductive track—the only track that leads to physical truth;—it utterly repudiates final causes, and thereby indicates a demoralized understanding on the part of its advocates. By the word, demoralized, I mean a want of capacity for comprehending the force of moral evidence, which is dependent on the highest faculties of our nature. . . . Are the highest conceptions of man, to which he is led by the necessities of his moral nature, to have no counterpart or fruition? Are they all a cheat and a mockery, and therefore out of harmony with nature? I say *no,* to all such questions; and fearlessly affirm that we cannot speculate on man's position in the actual world of nature, on his destinies, or *on his origin,* while we keep his highest faculties out of our sight. Strip him of these faculties, and he becomes entirely bestial; and he may well be (under such a false and narrow view) nothing better than the natural progeny of a beast, which has to live, to beget its likeness, and then die for ever.

By gazing only on material nature, a man may easily have his very senses bewildered . . . . he may become so frozen up, by a too long continued and exclusively material study, as to lose his relish for moral truth, and vivacity in apprehending it. I think I can see traces of this effect, both in the origin and in the details of certain portions of Darwin's theory; and, in confirmation of what I now write, I would appeal to all that he states about those marvellous structures,—the comb of a common honey-bee, and the eye of a mammal. His explanation of the phenomena—*viz.*, the perfection of the structures, and their adaptation to their office. There *is* a light by which a man may see and comprehend facts and truths such as these. But Darwin wilfully shuts it out from our senses; either because he does not apprehend its power, or because he disbelieves in its existence. This is the grand blemish of his work. Separated from his sterile and contracted theory, it contains very admirable details and beautiful views of nature—especially in those chapters which relate to the battle of life, the variations of species, and their diffusion through wide regions of the earth.

In some rare instances, Darwin shows a wonderful credulity. He seems to believe that a white bear, by being confined to the slops floating in the Polar basin, might in time be turned into a whale; that a lemur might easily be turned into a bat; that a three-toed tapir might be the great grandfather of a horse; or that the progeny of a horse (in America) have gone back into the tapir.

But any startling and (supposed) novel paradox—maintained very boldly and with an imposing plausibility, derived from a great array of facts all interpreted hypothetically—produces, in some minds, a kind of pleasing excitement, which predisposes them in its favor: and if they are unused to careful reflection, and averse to the labor of accurate investigation, they will be likely to conclude that what is (apparently) *original,* must be a production of original *genius,* and that anything very much opposed to prevailing notions must be a grand *discovery*. . . .

T. H. Huxley
ORTHODOXY SCOTCHED, IF NOT SLAIN

Thomas Henry Huxley (1825–1895) was called "Darwin's Bulldog" because of his almost endless fights and skirmishes in defense of evolution. Yet Huxley was much more than a defender of Darwin—as indeed is evident in this reading from the former's review of The Origin of Species *in 1860. As he said in his brief autobiography, Huxley's aim was to push forward "the application of scientific methods of investigation to all the problems of life." He thus became the great philosopher and popularizer of science and the scientific method in the nineteenth century. And for the purpose of championing this "New Reformation" of thought, Huxley opposed Bishop Wilberforce, Adam Sedgwick and anyone else of public stature who appeared to muzzle "veracity of thought and action" in the name of tradition or the "ecclesiastical spirit."*

Mr. Darwin's long-standing and well-earned scientific eminence probably renders him indifferent to that social notoriety which passes by the name of success; but if the calm spirit of the philosopher have not yet wholly superseded the ambition and the vanity of the carnal man within him, he must be well satisfied with the results of his venture in publishing the *Origin of Species.* Overflowing the narrow bounds of purely scientific circles, the "species question" divides with Italy and the Volunteers the attention of general society. Everybody has read Mr. Darwin's book, or, at least, has given an opinion upon its merits or demerits; pietists, whether lay or ecclesiastic, decry it with the mild railing which sounds so charitable; bigots denounce it with ignorant invective; old ladies of both sexes consider it a decidedly dangerous book, and even savants, who have no better mud to throw, quote antiquated writers to show that its author is no better than an ape himself; while every philosophical thinker hails it as a veritable Whitworth gun in the armoury of liberalism; and all competent naturalists and physiologists, whatever their opinions as to the ultimate fate of the doctrines put forth, acknowledge that the work in which they are embodied is a solid contribution to knowledge and inaugurates a new epoch in natural history.

Nor has the discussion of the subject been restrained within the

Abridged from Thomas Henry Huxley, *Lay Sermons, Addresses, and Reviews* (New York: D. Appleton, 1871).

FIGURES 3 and 4. Bishop Samuel Wilberforce (left) and T. H. Huxley (right) clashed over the implications of evolution. (*Vanity Fair, 1881*; photographs by Richard V. T. Stearns)

limits of conversation. When the public is eager and interested, reviewers must minister to its wants; and the genuine *littérateur* is too much in the habit of acquiring his knowledge from the book he judges—as the Abyssinian is said to provide himself with steaks from the ox which carries him—to be withheld from criticism of a profound scientific work by the mere want of the requisite preliminary scientific acquirement; while, on the other hand, the men of science who wish well to the new views, no less than those who dispute their validity, have naturally sought opportunities of expressing their opinions. Hence it is not surprising that almost all the critical journals have noticed Mr. Darwin's work at greater or less length; and so many disquisitions, of every degree of excellence, from the poor product of ignorance, too often stimulated by prejudice, to the fair and thoughtful essay of the candid student of Nature, have appeared, that it seems an almost helpless task to attempt to say anything new upon the question.

But it may be doubted if the knowledge and acumen of prejudged scientific opponents, or the subtlety of orthodox special pleaders, have yet exerted their full force in mystifying the real issues of the great controversy which has been set afoot, and whose end is hardly likely to be seen by this generation; so that at this eleventh hour, and even failing anything new, it may be useful to state afresh that which is true, and to put the fundamental positions advocated by Mr. Darwin in such a form that they may be grasped by those whose special studies lie in other directions. And the adoption of this course may be the more advisable, because notwithstanding its great deserts, and indeed partly on account of them, the *Origin of Species* is by no means an easy book to read—if by reading is implied the full comprehension of an author's meaning.

We do not speak jestingly in saying that it is Mr. Darwin's misfortune to know more about the question he has taken up than any man living. Personally and practically exercised in zoology, in minute anatomy, in geology; a student of geographical distribution, not on maps and in museums only, but by long voyages and laborious collection; having largely advanced each of these branches of science, and having spent many years in gathering and sifting materials for his present work, the store of accurately registered facts upon which the author of the *Origin of Species* is able to draw at will is prodigious.

But this very superabundance of matter must have been embarrassing to a writer who, for the present, can only put forward an abstract of his views; and thence it arises, perhaps, that notwithstanding the clearness of the style, those who attempt fairly to digest the book find much of it a sort of intellectual pemmican—a mass of facts crushed and pounded into shape, rather than held together by the ordinary medium of an obvious logical bond: due attention will, without doubt, discover this bond, but it is often hard to find.

* * *

Whatever may be his theoretical views, no naturalist will probably be disposed to demur to the following summary of that exposition:

Living beings, whether animals or plants, are divisible into multitudes of distinctly definable kinds, which are morphological species. They are also divisible into groups of individuals, which breed freely together, tending to reproduce their like, and are physiological species. Normally resembling their parents, the offspring of members of these species are still liable to vary, and the variation may be perpetuated by selection, as a race, which race, in many cases, presents all the characteristics of a morphological species. But it is not as yet proved that a race ever exhibits, when crossed with another race of the same species, those phenomena of hybridization which are exhibited by many species when crossed with other species. On the other hand, not only is it not proved that all species give rise to hybrids infertile *inter se,* but there is much reason to believe that, in crossing, species exhibit every gradation from perfect sterility to perfect fertility.

Such are the most essential characteristics of species. Even were man not one of them—a member of the same system and subject to the same laws—the question of their origin, their causal connection, that is, with the other phenomena of the universe, must have attracted his attention, as soon as his intelligence had raised itself above the level of his daily wants.

Indeed history relates that such was the case, and has embalmed for us the speculations upon the origin of living beings, which were among the earliest products of the dawning intellectual activity of man. In those early days positive knowledge was not to be had, but the craving after it needed, at all hazards, to be satisfied, and according to the country, or the turn of thought of the speculator, the suggestion that all living things arose from the mud of the Nile, from a primeval egg, or from some more anthropomorphic agency, afforded a sufficient resting-place for his curiosity. The myths of Paganism are as dead as Osiris or Zeus, and the man who should revive them, in opposition to the knowledge of our time, would be justly laughed to scorn; but the coeval imaginations current among the rude inhabitants of Palestine, recorded by writers whose very name and age are admitted by every scholar to be unknown, have unfortunately not yet shared their fate, but, even at this day, are regarded by nine-tenths of the civilized world as the authoritative standard of fact and the criterion of the justice of scientific conclusions, in all that relates to the origin of things, and, among them, of species. In this nineteenth century, as at the dawn of modern physical science, the cosmogony of the semi-barbarous Hebrew is the incubus of the philosopher and the opprobrium of the orthodox. Who shall number the patient and earnest seekers after truth, from the days of Galileo until now, whose lives have been embittered and their good name blasted by the mistaken zeal of Bibliolaters? Who shall count the host of weaker men whose sense of truth has been destroyed in the effort to harmonize impossibilities—whose life has been wasted in the attempt to force the generous new wine of Science into the old bottles of Judaism, compelled by the outcry of the same strong party?

It is true that if philosophers have suffered, their cause has been amply avenged. Extinguished theologians lie about the cradle of every science as the strangled snakes beside that of Hercules; and history records that whenever science and orthodoxy have been fairly opposed, the latter has been forced to retire from the lists, bleeding and crushed, if not annihilated; scotched, if not slain. But orthodoxy is the Bourbon of the world of thought. It learns not, neither can it forget; and though, at present, bewildered and afraid to move, it is as willing as ever to insist that the first chapter of Genesis contains the beginning and the end of sound science; and to visit, with such petty thunderbolts as its half-paralyzed hands can hurl, those who refuse to degrade Nature to the level of primitive Judaism.

Philosophers, on the other hand, have no such aggressive tendencies. With eyes fixed on the noble goal to which *per aspera et ardua* they tend, they may, now and then, be stirred to momentary wrath by the unnecessary obstacles with which the ignorant, or the malicious, encumber, if they cannot bar, the difficult path; but why should their souls be deeply vexed? The majesty of Fact is on their side, and the elemental forces of Nature are working for them. Not a star comes to the meridian at its calculated time but testifies to the justice of their methods—their beliefs are "one with the falling rain and with the growing corn." By doubt they are established, and open inquiry is their bosom friend. Such men have no fear of traditions however venerable, and no respect for them when they become mischievous and obstructive; but they have better than mere antiquarian business in hand, and if dogmas, which ought to be fossil but are not, are not forced upon their notice, they are too happy to treat them as nonexistent.

The hypotheses respecting the origin of species which profess to stand upon a scientific basis, and, as such, alone demand serious attention, are of two kinds. The one, the "special creation" hypothesis, presumes every species to have originated from one or more stocks, these not being the result of the modification of any other form of living matter—or arising by natural agencies—but being produced, as such, by a supernatural creative act.

The other, the so-called "transmutation" hypothesis, considers that all existing species are the result of the modification of pre-existing species, and those of their predecessors, by agencies similar to those which at the present day produce varieties and races, and therefore in an altogether natural way; and it is a probable, though not a necessary consequence of this hypothesis, that all living beings have arisen from a single stock. With respect to the origin of this primitive stock, or stocks, the doctrine of the origin of species is obviously not necessarily concerned. The transmutation hypothesis, for example, is perfectly consistent either with the conception of a special creation of the primitive germ, or with the supposition of its having arisen, as a modification of inorganic matter, by natural causes.

The doctrine of special creation owes its existence very largely to the supposed necessity of making science accord with the Hebrew cosmogony; but it is curious to observe that, as the doctrine is at present maintained by men of science, it is as hopelessly inconsistent with the Hebrew view as any other hypothesis.

If there be any result which has come more clearly out of geological investigation than another, it is, that the vast series of extinct animals and plants is not divisible, as it was once supposed to be, into distinct groups, separated by sharply marked boundaries. There are no great gulfs between epochs and formations—no successive periods marked by the appearance of plants, of water animals, and of land animals, en masse. Every year adds to the list of links between what the older geologists supposed to be widely separated epochs: witness the crags linking the drift with the older tertiaries; the Maestricht beds linking the tertiaries with the chalk; the St. Cassian beds exhibiting an abundant fauna of mixed mesozoic and paleozoic types, in rocks of an epoch once supposed to be eminently poor in life; witness, lastly, the incessant disputes as to whether a given stratum shall be reckoned devonian or carboniferous, silurian or devonian, cambrian or silurian.

This truth is further illustrated in a most interesting manner by the impartial and highly competent testimony of M. Pictet, from whose calculations of what percentage of the genera of animals, existing in any formation, lived during the preceding formation, it results that in no case is the proportion less than *one-third,* or 33 per cent. It is the triassic formation, or the commencement of the mesozoic epoch, which has received this smallest inheritance from preceding ages. The other formations not uncommonly exhibit 60, 80, or even 94 per cent of genera in common with those whose remains are imbedded in their predecessor. Not only is this true, but the subdivisions of each formation exhibit new species characteristic of, and found only in, them; and, in many cases, as in the lias for example, the separate beds of these subdivisions are distinguished by well-marked and peculiar forms of life. A section, a hundred feet thick, will exhibit, at different heights, a dozen species of ammonite, none of which passes beyond its particular zone of limestone, or clay, into the zone below it or into that above it; so that those who adopt the doctrine of special creation must be prepared to admit, that at intervals of time, corresponding with the thickness of these beds, the Creator thought fit to interfere with the natural course of events for the purpose of making a new ammonite. It is not easy to transplant oneself into the frame of mind of those who can accept such a conclusion as this, on any evidence short of absolute demonstration; and it is difficult to see what is to be gained by so doing, since, as we have said, it is obvious that such a view of the origin of living beings is utterly opposed to the Hebrew cosmogony. Deserving no aid from the powerful arm of bibliolatry, then, does the received form of the hypothesis of special creation derive any support from science or sound logic? Assuredly not much.

But the hypothesis of special creation is not only a mere specious mask for our ignorance; its existence in Biology marks the youth and imperfection of the science. For what is the history of every science but the history of the elimination of the notion of creative, or other interferences, with the natural order of the phenomena which are the subject-matter of that science? When Astronomy was young "the morning stars sang together for joy," and the planets were guided in their courses by celestial hands. Now, the harmony of the stars has resolved itself into gravitation according to the inverse squares of the distances, and the orbits of the planets are deducible from the laws of the forces which allow a schoolboy's stone to break a window. The lightning was the angel of the Lord; but it has pleased Providence, in these modern times, that science should make it the humble messenger of man, and we know that every flash that shimmers about the horizon on a summer's evening is determined by ascertainable conditions, and that its direction and brightness might, if our knowledge of these were great enough, have been calculated.

Harmonious order governing eternally continuous progress—the web and woof of matter and force interweaving by slow degrees, without a broken thread, that veil which lies between us and the Infinite—that universe which alone we know or can know; such is the picture which science draws of the world, and in proportion as any part of that picture is in unison with the rest, so may we feel sure that it is rightly painted. Shall Biology alone remain out of harmony with her sister sciences?

Two years ago, in fact, though we venture to question if even the strongest supporters of the special creation hypothesis had not, now and then, an uneasy consciousness that all was not right, their position seemed more impregnable than ever, if not by its own inherent strength, at any rate by the obvious failure of all the attempts which had been made to carry it. On the other hand, however much the few, who thought deeply on the question of species, might be repelled by the generally received dogmas, they saw no way of escaping from them, save by the adoption of suppositions, so little justified by experiment or by observation, as to be at least equally distasteful.

Such being the general ferment in the minds of naturalists, it is no wonder that they mustered strong in the rooms of the Linnaean Society, on the 1st of July of the year 1858, to hear two papers by authors living on opposite sides of the globe, working out their results independently, and yet professing to have discovered one and the same solution of all the problems connected with species. The one of these authors was an able naturalist, Mr. Wallace, who had been employed for some years in studying the productions of the islands of the Indian Archipelago, and who had forwarded a memoir embodying his views to Mr. Darwin, for communication to the Linnaean Society. On perusing the essay, Mr. Darwin was not a little surprised to find that it embodied some of the leading ideas of a great work which he had been preparing for twenty years, and parts of which, containing a development of the very same views, had been perused by his private friends fifteen or sixteen years before. Perplexed in what manner to do full justice both to his friend and to himself, Mr. Darwin placed the matter in the hands of Dr. Hooker and Sir Charles Lyell, by whose advice he communicated a brief abstract of his own views to the Linnaean Society, at the same time that Mr. Wallace's paper was read. Of that abstract, the work on the *Origin of Species* is an enlargement; but a complete statement of Mr. Darwin's doctrine is looked for in the large and well-illustrated work which he is said to be preparing for publication.

The Darwinian hypothesis has the merit of being eminently simple and comprehensible in principle, and its essential positions may be stated in a very few words: all species have been produced by the development of varieties from common stocks by the conversion of these first into permanent races and then into new species, by the process of *natural selection*, which process is essentially identical with that artificial selection by which man has originated the races of domestic animals—the *struggle for existence* taking the place of man, and exerting, in the case of natural selection, that selective action which he performs in artificial selection.

There cannot be a doubt that the method of inquiry which Mr. Darwin has adopted is not only rigorously in accordance with the canons of scientific logic, but that it is the only adequate method. Critics exclusively trained in classics or in mathematics, who have never determined a scientific fact in their lives by induction from experiment or observation, prate learnedly about Mr. Darwin's method, which is not inductive enough, not Baconian enough, forsooth, for them.

. . . Inductively, Mr. Darwin endeavors to prove that species arise in a given way. Deductively, he desires to show that, if they arise in that way, the facts of distribution, development, classification, etc., may be accounted for, i.e. may be deduced from their mode of origin, combined with admitted changes in physical geography and climate, during an indefinite period. And this explanation, or coincidence of observed with deduced facts, is, so far as it extends, a verification of the Darwinian view.

After much consideration, and with assuredly no bias against Mr. Darwin's views, it is our clear conviction that, as the evidence stands, it is not absolutely proven that a group of animals, having all the characters exhibited by species in Nature, has ever been originated by selection, whether artificial or natural. Groups having the morphological character of species, distinct and permanent races in fact, have been so produced over and over again; but there is no positive evidence, at present, that any group of animals has, by variation and selective breeding, given rise to another group which was even in the least degree infertile with the first. Mr. Darwin is perfectly aware of this weak point, and brings forward a multitude of ingenious and important arguments to diminish the force of the objection. We admit the value of these arguments to their fullest extent; nay, we will go so far as to express our belief that experiments, conducted by a skilful physiologist, would very probably obtain the desired production of mutually more or less infertile breeds from a common stock, in a comparatively few years; but still, as the case stands at present, this "little rift within the lute" is not to be disguised nor overlooked.

But we must pause. The discussion of Mr. Darwin's arguments in detail would lead us far beyond the limits within which we proposed, at starting, to confine this article. Our object has been attained if we have given an intelligible, however brief, account of the established facts connected with species, and of the relation of the explanation of those facts offered by Mr. Darwin to the theoretical views held by his predecessors and his contemporaries, and, above all, to the requirements of scientific logic. We have ventured to point

out that it does not, as yet, satisfy all those requirements; but we do not hesitate to assert that it is as superior to any preceding or contemporary hypothesis, in the extent of observational and experimental basis on which it rests, in its rigorously scientific method, and in its power of explaining biological phenomena, as was the hypothesis of Copernicus to the speculations of Ptolemy. But the planetary orbits turned out to be not quite circular after all, and, grand as was the service Copernicus rendered to science, Kepler and Newton had to come after him. What if the orbit of Darwinism should be a little too circular? What if species should offer residual phenomena, here and there, not explicable by natural selection? Twenty years hence naturalists may be in a position to say whether this is, or is not, the case; but in either event they will owe the author of *The Origin of Species* an immense debt of gratitude. . . . And viewed as a whole, we do not believe that, since the publication of Von Baer's *Researches on Development,* thirty years ago, any work has appeared calculated to exert so large an influence, not only on the future of Biology, but in extending the domination of Science over regions of thought into which she has, as yet, hardly penetrated.

FIGURE 5. The liberal New York magazine *Puck* depicted Darwin as "A Sun of the 19th Century," causing clouds of bigotry and superstition to vanish.

Leslie Stephen
SPREADING AGNOSTICISM

Shortly after T. H. Huxley invented the term agnosticism in 1869, Leslie Stephen (1832–1904) began to call himself an agnostic and defend and popularize the ideas that it symbolized. Agnosticism (a-gnosis, or without knowledge) summarized the stance of literati like T. H. Huxley and Leslie Stephen, for whom belief in God or some specific definition of ultimate reality lacked any proof. Stephen had been ordained as an Anglican priest in 1859, but lost his faith not long after reading Darwin's Origin of Species that same year. In the doubt-inflicted decades of the 1860s and 1870s Stephen challenged his English and American readers to realize that traditional religious and philosophical beliefs were illusionary, hopelessly irrelevant and contradictory, and morally incompatible with the evil and injustice in the world. This reading from An Agnostic's Apology (first published in 1876) displays these opinions and shows how Stephen discovered in science a partial guide through the "mists and darkness" of life, and found in technology a secular security that for him rendered many religious activities passé.

The name Agnostic, originally coined by Professor Huxley about 1869, has gained general acceptance. . . . The old theological phrase for an intellectual opponent was Atheist—a name which still retains a certain flavor as of the stake in this world and hell-fire in the next, and which, moreover, implies an inaccuracy of some importance. Dogmatic Atheism—the doctrine that there is no God, whatever may be meant by God—is, to say the least, a rare phase of opinion. The word Agnosticism, on the other hand, seems to imply a fairly accurate appreciation of a form of creed already common and daily spreading. The Agnostic is one who asserts—what no one denies—that there are limits to the sphere of human intelligence. He asserts, further . . . that those limits are such as to exclude . . . "metempirical" knowledge . . . and asserts . . . that theology lies within this forbidden sphere. This last assertion raises the important issue; and, though I have no pretension to invent an opposition nickname, I may venture, for the purposes of this article, to describe the rival school as Gnostics.

The Gnostic holds that our reason can, in some sense, transcend

Abridged from Leslie Stephen, *An Agnostic's Apology* (New York: G. P. Putnam's Sons, 1903 [1876]).

the narrow limits of experience. He holds that we can attain truths not capable of verification, and not needing verification, by actual experiment or observation. He holds, further, that a knowledge of those truths is essential to the highest interests of mankind, and enables us in some sort to solve the dark riddle of the universe. . . . This knowledge is embodied in the central dogma of theology. God is the name of the harmony; and God is knowable. Who would not be happy in accepting this belief, if he could accept it honestly? Who would not be glad if he could say with confidence: "the evil is transitory, the good eternal: our doubts are due to limitations destined to be abolished, and the world is really an embodiment of love and wisdom, however dark it may appear to our faculties"? And yet, if the so-called knowledge be illusory, are we not bound by the most sacred obligations to recognize the facts? . . . Dreams may be pleasanter for the moment than realities; but happiness must be won by adapting our lives to the realities. And who, that has felt the burden of existence, and suffered under well-meant efforts at consolation, will deny that such consolations are the bitterest of mockeries?

Besides the important question whether the Gnostic can prove his dogmas, there is, therefore, the further question whether the dogmas, if granted, have any meaning. Do they answer our doubts, or mock us with the appearance of an answer? The Gnostics rejoice in their knowledge. Have they anything to tell us?

Not long ago, at least, there appeared in the papers a string of propositions framed—so we were assured—by some of the most candid and most learned of living theologians. These propositions defined by the help of various languages the precise relations which exist between the persons of the Trinity. . . . It is enough to say that they defined the nature of God Almighty with an accuracy from which modest naturalists would shrink in describing the genesis of a black-beetle. I know not whether these dogmas were put forward as articles of faith, as pious conjectures, or as tentative contributions to a sound theory. At any rate, it was supposed that they were interesting to beings of flesh and blood. If so, one can only ask in wonder whether an utter want of reverence is most strongly implied in this mode of dealing with sacred mysteries; or an utter ignorance of the existing state of the world in the assumption that the question which really divides mankind is the double procession of the Holy Ghost; or an utter incapacity for speculation in the confusion of these dead exuviae of long-past modes of thought with living intellectual tissue; or an utter want of imagination, or of even a rudimentary sense of humor, in the hypothesis that the promulgation of such dogmas could produce anything but the laughter of sceptics and the contempt of the healthy human intellect?

We are struggling with hard facts, and they would arm us with the forgotten tools of scholasticism. We wish for spiritual food, and are to be put off with these ancient mummeries of forgotten dogma. If Agnosticism is the frame of mind which summarily rejects these imbecilities, and would restrain the human intellect from wasting its powers on the attempt to galvanize into sham activity this *caput mortuum* of old theology, nobody need be afraid of the name. Argument against such adversaries would be itself a foolish waste of time. Let the dead bury their dead, and Old Catholics decide whether the Holy Ghost proceeds from the Father and the Son, or from the Father alone. Gentlemen, indeed, who still read the Athanasian Creed, and profess to attach some meaning to its statements, have no right to sneer at their brethren who persist in taking things seriously. But for men who long for facts instead of phrases, the only possible course is to allow such vagaries to take their own course to the limbo to which they are naturally destined.

You tell us to be ashamed of professing ignorance. Where is the shame of ignorance in matters still involved in endless and hopeless controversy? Is it not rather a duty? Why should a lad who has just run the gauntlet of examinations and escaped to a country parsonage be dogmatic, when his dogmas are denounced as erroneous by half the philosophers of the world? What theory of the universe am I to accept as demonstrably established? At the very earliest dawn of philosophy men were divided by earlier forms of the same problems which divide them now. Shall I be a Platonist or an Aristotelian? Shall I admit or deny the existence of innate ideas? Shall I believe in the possibility or in the impossibility of transcending experience? ... Can I stop where Descartes stopped, or must I go on to Spinoza? Or shall I follow Locke's guidance, and end with Hume's scepticism? Or listen to Kant, and, if so, shall I decide that he is right in destroying theology, or in reconstructing it, or in both performances? Does Hegel hold the key of the secret, or is he a mere spinner of jargon? May not Feuerbach or Schopenhauer represent the true development of metaphysical inquiry? . . . State any one proposition in which all philosophers agree, and I will admit it to be true; or any one which has a manifest balance of authority, and I will agree that it is probable. But so long as every philosopher flatly contradicts the first principles of his predecessors, why affect certainty? The only agreement I can discover is, that there is no philosopher of whom his opponents have not said that his opinions lead logically either to Pantheism or to Atheism.

When all the witnesses thus contradict each other, the *prima facie* result is pure scepticism. There is no certainty. Who am I, if I were the ablest of modern thinkers, to say summarily that all the great men who differed from me are wrong, and so wrong that their difference should not even raise a doubt in my mind? From such scepticism there is indeed one, and, so far as I can see, but one, escape. The very hopelessness of the controversy shows that the reasoners have been transcending the limits of reason. They have reached a point where, as at the pole, the compass points indifferently to every quarter. Thus there is a chance that I may retain what is valuable in the chaos of speculation, and reject what is bewildering by confining the mind to its proper limits. But has any limit ever been suggested, except a limit which comes in substance to an exclusion of all ontology? In short, if I would avoid utter scepticism, must I not be an Agnostic?

The ancient difficulty which has perplexed men since the days of Job is this: Why are happiness and misery arbitrarily distributed? Why do the good so often suffer, and the evil so often flourish? The difficulty, says the determinist, arises entirely from applying the conception of justice where it is manifestly out of place. The advocate of free-will refuses this escape, and is perplexed by a further difficulty. Why are virtue and vice arbitrarily distributed? Of all the puzzles of this dark world, or of all forms of the one great puzzle, the most appalling is that which meets us at the corner of every street. Look at the children growing up amidst moral poison; see the brothel and the public-house turning out harlots and drunkards by the thousand; at the brutalized elders preaching cruelty and shamelessness by example; and deny, if you can, that lust and brutality are generated as certainly as scrofula and typhus. Nobody dares to deny it. All philanthropists admit it; and every hope of improvement is based on the assumption that the moral character is determined by its surroundings. What does the theological advocate of free-will say to reconcile such a spectacle with our moral conceptions? Will God damn all these wretches for faults due to causes as much beyond their power as the shape of their limbs or as the orbits of the planets? Or will He make some allowance, and decline to ask for grapes from thistles, and exact purity of life from beings born in corruption, breathing corruption, and trained in corruption?

For anything we can tell—for we know nothing of the circumstances of their birth and education—the effort which Judas Iscariot exerted in restoring the price of blood may have required a greater force of free-will than would have saved Peter from denying his Master. Moll Flanders may put forth more power to keep out of the lowest depths of vice than a girl brought up in a convent to kill herself by ascetic austerities. If, in short, reward is proportioned to virtue, it cannot be proportioned to merit, for merit, by the hypothesis, is proportioned to the free-will, which is only one of the factors of virtue. The apparent injustice may, of course, be remedied by some unknowable compensation; but for all that appears, it is the height of injustice to reward equally equal attainments under entirely different conditions. In other words, the theologian has raised a difficulty from which he can only escape by the help of Agnosticism. Justice is not to be found in the visible arrangements of the universe.

This world . . . is a chaos, in which the most conspicuous fact is the absence of the Creator. Nay, it is so chaotic that, according to theologians, infinite rewards and penalties are required to square the account and redress the injustice here accumulated.

* * *

Is happiness a dream, or misery, or is it all a dream? Does not our answer vary with our health and with our condition? When, rapt in the security of a happy life, we cannot even conceive that our happiness will fail, we are practical optimists. When some random blow out of the dark crushes the pillars round which our life has been entwined as recklessly as a boy sweeps away a cobweb, when at a single step we plunge through the flimsy crust of happiness into the deep gulfs beneath, we are tempted to turn to Pessimism. Who shall decide, and how? Of all questions that can be asked, the most important is surely this: Is the tangled web of this world composed chiefly of happiness or of misery? And of all questions that can be asked, it is surely the most unanswerable. For in no other problem is the difficulty of discarding the illusions arising from our own experience, of eliminating "the personal error" and gaining an outside standing-point, so hopeless.

In any case the real appeal must be to experience. Ontologists may manufacture libraries of jargon without touching the point. They have never made, or suggested the barest possibility of making, a bridge from the world of pure reason to the contingent world in which we live.

What, then, is the net result? One insoluble doubt has haunted men's minds since thought began in the world. No answer has ever been suggested. One school of philosophers hands it to the next. It is denied in one form only to reappear in another. The question is not which system excludes the doubt, but how it expresses the doubt. Admit or deny the competence of reason in theory, we all agree that it fails in practice. Theologians revile reason as much as Agnostics; they then appeal to it, and it decides against them. They amend their plea by excluding certain questions from its jurisdiction, and those questions include the whole difficulty. They go to revelation, and revelation replies by calling doubt, mystery. They declare that their consciousness declares just what they want it to declare. Ours declares something else. Who is to decide? The only appeal is to experience, and to appeal to experience is to admit the fundamental dogma of Agnosticism.

Is it not, then, the very height of audacity, in face of a difficulty which meets us at every turn, which has perplexed all the ablest thinkers in proportion to their ability, which vanishes in one shape only to show itself in another, to declare roundly, not only that the difficulty can be solved, but that it does not exist? Why, when no honest man will deny in private that every ultimate problem is wrapped in the profoundest mystery, do honest men proclaim in pulpits that unhesitating certainty is the duty of the most foolish and ignorant? Is it not a spectacle to make the angels laugh? We are a company of ignorant beings, feeling our way through mists and darkness, learning only by incessantly repeated blunders, obtaining a glimmering of truth by falling into every conceivable error, dimly discerning light enough for our daily needs, but hopelessly differing whenever we attempt to describe the ultimate origin or end of our paths; and yet, when one of us ventures to declare that we don't know the map of the universe as well as the map of our infinitesimal parish, he is hooted, reviled, and perhaps told that he will be damned to all eternity for his faithlessness. Amidst all the endless and hopeless controversies which have left nothing but bare husks of meaningless words, we have been able to discover certain reliable truths. They don't take us very far, and the condition of discovering them has been distrust of a priori guesses, and the systematic interrogation of experience. Let us, say some of us, follow at least this clue. Here we shall find sufficient guidance for the needs of life, though we renounce forever the attempt to get behind the veil which no one has succeeded in raising; if, indeed, there be anything behind. You miserable Agnostics! is the retort; throw aside such rubbish, and cling to the old husks. Stick to the words which profess to explain everything; call your doubts mysteries, and they won't disturb you any longer; and believe in those necessary truths of which no two philosophers have ever succeeded in giving the same version.

Gentlemen, we can only reply, wait till you have some show of agreement amongst yourselves. Wait till you can give some answer, not palpably a verbal answer, to some one of the doubts which oppress us as they oppress you. Wait till you can point to some single truth, however trifling, which has been discovered by your method, and will stand the test of discussion and verification. Wait till you can appeal to reason without in the same breath vilifying reason. Wait till your Divine revelations have something more to reveal than the hope that the hideous doubts which they suggest may possibly be without foundation. Till then we shall be content to admit openly, what you whisper under your breath or hide in technical jargon, that the ancient secret is a secret still; that man knows nothing of the Infinite and Absolute; and that, knowing nothing, he had better not be dogmatic about his ignorance.

What is to be the religion of the future? I have not the slightest idea. I at least am perfectly certain of my own ignorance, and I have a strong impression that almost everyone else is equally ignorant. I can see, as everyone else can see, that a vast social and intellectual transformation is taking place—and taking place, probably, with more rapidity now than at almost any historical period. I can dimly

Spreading Agnosticism

guess at some of the main characteristics of the process. I can discover some conditions, both of the social and the speculative kind, which will probably influence the result. I cannot doubt that some ancient doctrines have lost their vitality, and that some new beliefs must be recognized by one who would influence the minds of the coming generations. I cannot believe in the simple resurrection of effete religious ideas; nor, on the other hand, do I believe that the ideas which still have life have as yet been effectually embodied in any system which professes to take the place of the old.

The argument of the more hopeful would be that, after all, modern science is what people call a "great fact." The existence of a vast body of definitively established truths, forming an organized and coherent system, giving proofs of its vitality by continuous growth, and of its ability by innumerable applications to our daily wants, is not only an important element in the question, but it is the most conspicuous point of difference between the purely intellectual conditions of the contemporary evolution and that which resulted in the triumph of Christianity. Here is the fixed fulcrum, an unassailable nucleus of definite belief, round which all other beliefs must crystallize. It supplies a ground, intelligible in some relations to the ordinary mind, upon which the philosopher may base his claims to respect. Whatever system would really prevail must be capable of assimilating modern scientific theories; for a direct assault is hopeless, and to ignore science is impossible. The enormous apologetic literature destined to reconcile faith and reason is a sufficient proof that the reconciliation is a necessity for the old faith—and that it is an impossibility. . . . Science, moreover, touches men's interests at so many points that it has the key of the position. . . . Science means steam-engines, telegraphy, and machinery, and, whether the reflection be consolatory or the reverse, we may be fully confident that all the power of all the priests and all the philosophers in the world would be as idle wind if directed against these palpable daily conveniences. And, undoubtedly, this consideration is enough to imply that scientific thought is a force to be taken into account. There are directions in which the incompatibility between its results and those of the old creeds is felt by ordinary minds. We still pray for a fine harvest; but we really consult the barometer, and believe more in the prophecies of the meteorologist than in an answer to our prayers; *Te Deums* for victories excite more ridicule than sympathy; and we encounter the cholera by improved systems of drainage, without attributing much value to fasting and processions.

The creed of the future, whatever it may be, exists only in germ. You cannot give to a believer anything in place of his cult, of the sacred symbols which reflect his emotions, of the whole system of disciplined and organized modes of worship, of prayers, of communion with his fellows, which to him are the great attraction of his religion. You cannot even tell him what system is likely to replace them hereafter, or whether human nature is so constituted that it will be able simply to drop the old without replacing it by anything directly analogous. And, therefore, you must admit that for the present a man who would abandon the old doctrines is compelled to stand alone. He must find sufficient comfort in the consciousness that he is dealing honestly with his intellect; he must be able to dispense with the old consolations of heaven and hell; he must be content to admit explicitly that the ancient secret has not been revealed, and to hold that people will be able to get on somehow or other, even if the most ignorant and stupid cease to consider it a solemn duty to dogmatize with the utmost confidence upon matters of which the wisest know absolutely nothing, and never expect to know anything. Undoubtedly, this is to accept a position from which many people will shrink; and it is pleasanter to the ordinary mind to reject it summarily as untenable, or to run up some temporary refuge of fine phrases, and try to believe in its permanence.

Asa Gray
THE COMPATIBILITY OF EVOLUTION AND RELIGION

The greatest advocate and popularizer of evolution in America was Asa Gray (1810–1888), a professor of natural science at Harvard and the outstanding American botanist of his time. Gray had exchanged ideas with Darwin for four years before the publication of The Origin of Species, *and had "confirmed" evolution by his own comparative study of Japanese and North American flora. Gray was thus well prepared to defend Darwinism against American traditionalists, not the least of whom was his Harvard colleague, Louis Agassiz (1807–1873), a disciple of Georges Cuvier and confessed creationalist.*

The defense of evolution involved for Gray both a lucid exposition of Darwin's biological insights and a constant demonstration of the compatibility of those insights with teleological design. Gray believed that Darwinian evolution gave a new conceptual basis for teleology, not a secularized substitute for the same. By offering a more thorough understanding of the history of marvelously designed organisms and of the methods used in producing them, evolutionary teleology, thought Gray, was far superior to the older, Paleyean point of view. This reading is drawn from two chapters of Gray's Darwiniana (1876), *the former chapter being first published as an article in the* Atlantic Monthly *in 1860.*

It is undeniable that Mr. Darwin has purposely been silent upon the philosophical and theological applications of his theory. . . . Here, as in higher instances, confident as we are that there is a final cause, we must not be over-confident that we can infer the particular or true one. Perhaps the author is more familiar with natural-historical than with philosophical inquiries, and, not having decided which particular theory about efficient cause is best founded, he meanwhile argues the scientific questions concerned—all that relates to secondary causes—upon purely scientific grounds, as he must do in any case. Perhaps, confident, as he evidently is, that his view will finally be adopted, he may enjoy a sort of satisfaction in hearing it denounced as sheer atheism by the inconsiderate, and afterward, when it takes its place with the nebular hypothesis and the like, see this judgment reversed, as we suppose it would be in such event.

Whatever Mr. Darwin's philosophy may be, or whether he has any,

Abridged from Asa Gray, *Darwiniana* (New York: D. Appleton, 1876).

is a matter of no consequence at all, compared with the important questions, whether a theory to account for the origination and diversification of animal and vegetable forms through the operation of secondary causes does or does not exclude design; and whether the establishment by adequate evidence of Darwin's particular theory of diversification through variation and natural selection would essentially alter the present scientific and philosophical grounds for theistic views of Nature. . . . After full and serious consideration, we are constrained to say that, in our opinion, the adoption of a derivative hypothesis, and of Darwin's particular hypothesis, if we understand it, would leave the doctrines of final causes, utility, and special design, just where they were before. We do not pretend that the subject is not environed with difficulties. Every view is so environed; and every shifting of the view is likely, if it removes some difficulties, to bring others into prominence. But we cannot perceive that Darwin's theory brings in any new kind of scientific difficulty, that is, any with which philosophical naturalists were not already familiar.

Wherefore, Darwin's reticence about efficient cause does not disturb us. He considers only the scientific questions. As already stated, we think that a theistic view of Nature is implied in his book, and we must charitably refrain from suggesting the contrary until the contrary is logically deduced from his premises.

* * *

If we believe that the species were designed, and that natural propagation was designed, how can we say that the actual varieties of the species were not equally designed? Have we not similar grounds for inferring design in the supposed varieties of species, that we have in the case of the supposed species of a genus? When a naturalist comes to regard as three closely related species what he before took to be so many varieties of one species, how has he thereby strengthened our conviction that the three forms are designed to have the differences which they actually exhibit? Wherefore, so long as gradatory, orderly, and adapted forms in Nature argue design, and at least while the physical cause of variation is utterly unknown and mysterious, we should advise Mr. Darwin to assume, in the philosophy of his hypothesis, that variation has been led along certain beneficial lines. Streams flowing over a sloping plain by gravitation (here the counterpart of natural selection) may have worn their actual chan-

The Compatibility of Evolution and Religion 115

nels as they flowed; yet their particular courses may have been assigned; and where we see them forming definite and useful lines of irrigation, after a manner unaccountable on the laws of gravitation and dynamics, we should believe that the distribution was designed.

To insist, therefore, that the new hypothesis of the derivative origin of the actual species is incompatible with final causes and design, is to take a position which we must consider philosophically untenable.

The whole argument in natural theology proceeds upon the ground that the inference for a final cause of the structure of the hand and of the valves in the veins is just as valid now, in individuals produced through natural generation, as it would have been in the case of the first man, supernaturally created. Why not, then, just as good even on the supposition of the descent of men from chimpanzees and gorillas, since those animals possess these same contrivances? Or, to take a more supposable case: If the argument from structure to design is convincing when drawn from a particular animal, say a Newfoundland dog, and is not weakened by the knowledge that this dog came from similar parents, would it be at all weakened if, in tracing his genealogy, it were ascertained that he was a remote descendant of the mastiff or some other breed, or that both these and other breeds came (as is suspected) from some wolf? If not, how is the argument for design in the structure of our particular dog affected by the supposition that his wolfish progenitor came from a post-tertiary wolf, perhaps less unlike an existing one than the dog in question is to some other of the numerous existing races of dogs, and that this post-tertiary came from an equally or more different tertiary wolf? And if the argument from structure to design is not invalidated by our present knowledge that our individual dog was developed from a single organic cell, how is it invalidated by the supposition of an analogous natural descent, through a long line of connected forms, from such a cell, or from some simple animal, existing ages before there were any dogs?

Again, suppose we have two well-known and apparently most decidedly different animals or plants, A and D, both presenting, in their structure and in their adaptations to the conditions of existence, as valid and clear evidence of design as any animal or plant ever presented: suppose we have now discovered two intermediate species, B and C, which make up a series with equable differences from A to

D. Is the proof of design or final cause in A and D, whatever it amounted to, at all weakened by the discovery of the intermediate forms? Rather does not the proof extend to the intermediate species, and go to show that all four were equally designed? Suppose, now, the number of intermediate forms to be much increased, and therefore the gradations to be closer yet—as close as those between the various sorts of dogs, or races of men, or of horned cattle: would the evidence of design, as shown in the structure of any of the members of the series, be any weaker than it was in the case of A and D? Whoever contends that it would be, should likewise maintain that the origination of individuals by generation is incompatible with design, or an impossibility in Nature. We might all have confidently thought the latter, antecedently to experience of the fact of reproduction. Let our experience teach us wisdom.

These illustrations make it clear that the evidence of design from structure and adaptation is furnished *complete* by the individual animal or plant itself, and that our knowledge or our ignorance of the history of its formation or mode of production adds nothing to it and takes nothing away. We infer design from certain arrangements and results; and we have no other way of ascertaining it. . . . Some arrangements in Nature appear to be contrivances, but may leave us in doubt. Many others, of which the eye and the hand are notable examples, compel belief with a force not appreciably short of demonstration. Clearly to settle that such as these must have been designed goes far toward proving that other organs and other seemingly less explicit adaptations in Nature must also have been designed, and clinches our belief . . . that all Nature is a preconcerted arrangement, a manifested design.

We could not affirm that the arguments for design in Nature are conclusive to all minds. But we may insist, upon grounds already intimated, that, whatever they were good for before Darwin's book appeared, they are good for now. To our minds the argument from design always appeared conclusive of the being and continued operation of an intelligent First Cause, the Ordainer of Nature; and we do not see that the grounds of such belief would be disturbed or shifted by the adoption of Darwin's hypothesis. We are not blind to the philosophical difficulties which the thoroughgoing implication of design in Nature has to encounter, nor is it our vocation to obviate them. It suffices us to know that they are not new nor peculiar diffi-

The Compatibility of Evolution and Religion

culties—that, as Darwin's theory and our reasonings upon it did not raise these perturbing spirits, they are not bound to lay them. Meanwhile, that the doctrine of design encounters the very same difficulties in the material that it does in the moral world is just what ought to be expected.

So the issue between the skeptic and the theist is only the old one, long ago argued out—namely, whether organic Nature is a result of design or of chance. Variation and natural selection open no third alternative; they concern only the question how the results, whether fortuitous or designed, may have been brought about. Organic Nature abounds with unmistakable and irresistible indications of design, and, being a connected and consistent system, this evidence carries the implication of design throughout the whole. On the other hand, chance carries no probabilities with it, can never be developed into a consistent system, but, when applied to the explanation of orderly or beneficial results, heaps up improbabilities at every step beyond all computation. To us, a fortuitous Cosmos is simply inconceivable. The alternative is a designed Cosmos.

It is very easy to assume that, because events in Nature are in one sense accidental, and the operative forces which bring them to pass are themselves blind and unintelligent (physically considered, all forces are), therefore they are undirected, or that he who describes these events as the results of such forces thereby assumes that they are undirected. This is the assumption of the Boston reviewers, and of Mr. Agassiz, who insists that the only alternative to the doctrine, that all organized beings were supernaturally created just as they are, is, that they have arisen *spontaneously* through the *omnipotence of matter.*

As to all this, nothing is easier than to bring out in the conclusion what you introduce in the premises. If you import atheism into your conception of variation and natural selection, you can readily exhibit it in the result. If you do not put it in, perhaps there need be none to come out.

* * *

It is evident that the strongest point against the compatibility of Darwin's hypothesis with design in Nature is made when natural selection is referred to as picking out those variations which are improvements from a vast number which are not improvements, but perhaps the contrary, and therefore useless or purposeless, and born to perish. But even here the difficulty is not peculiar; for Nature abounds with analogous instances. Some of our race are useless, or worse, as regards the improvement of mankind; yet the race may be designed to improve, and may be actually improving. Or, to avoid the complication with free agency—the whole animate life of a country depends absolutely upon the vegetation, the vegetation upon the rain. The moisture is furnished by the ocean, is raised by the sun's heat from the ocean's surface, and is wafted inland by the winds. But what multitudes of raindrops fall back into the ocean— are as much without a final cause as the incipient varieties which come to nothing! Does it therefore follow that the rains which are bestowed upon the soil with such rule and average regularity were not designed to support vegetable and animal life? Consider, likewise, the vast proportion of seeds and pollen, of ova and young—a thousand or more to one—which come to nothing, and are therefore purposeless in the same sense, and only in the same sense, as are Darwin's unimproved and unused slight variations. The world is full of such cases; and these must answer the argument—for we cannot, except by thus showing that it proves too much.

Finally, it is worth noticing that, though natural selection is scientifically explicable, variation is not. Thus far the cause of variation, or the reason why the offspring is sometimes unlike the parents, is just as mysterious as the reason why it is generally like the parents. It is now as inexplicable as any other origination; and, if ever explained, the explanation will only carry up the sequence of secondary causes one step farther, and bring us in face of a somewhat different problem, but which will have the same element of mystery that the problem of variation has now. Circumstances may preserve or may destroy the variations; man may use or direct them; but selection, whether artificial or natural, no more originates them than man originates the power which turns a wheel, when he dams a stream and lets the water fall upon it. The origination of this power is a question about efficient cause. The tendency of science in respect to this obviously is not toward the omnipotence of matter, as some suppose, but toward the omnipotence of spirit.

So the real question we come to is as to the way in which we are to conceive intelligent and efficient cause to be exerted, and upon what exerted. Are we bound to suppose efficient cause in all cases

The Compatibility of Evolution and Religion

exerted upon nothing to evoke something into existence—and this thousands of times repeated, when a slight change in the details would make all the difference between successive species? Why may not the new species, or some of them, be designed diversifications of the old?

There are, perhaps, only three views of efficient cause which may claim to be both philosophical and theistic:

1. The view of its exertion at the beginning of time, endowing matter and created things with forces which do the work and produce the phenomena.
2. This same view, with the theory of insulated interpositions, or occasional direct action, engrafted upon it—the view that events and operations in general go on in virtue simply of forces communicated at the first, but that now and then, and only now and then, the Deity puts his hand directly to the work.
3. The theory of the immediate, orderly, and constant, however infinitely diversified, action of the intelligent efficient Cause.

It must be allowed that, while the third is preeminently the Christian view, all three are philosophically compatible with design in Nature. The second is probably the popular conception. Perhaps most thoughtful people oscillate from the middle view toward the first or the third—adopting the first on some occasions, the third on others.

* * *

The combination of the principle of design with the hypothesis of the immutability and isolated creation of species . . . is at present a hindrance rather than a help to any just and consistent teleology.

By the adoption of the Darwinian hypothesis, or something like it, which we incline to favor, many . . . difficulties are obviated, and others diminished. In the comprehensive and far-reaching teleology which may take the place of the former narrow conceptions, organs and even faculties, useless to the individual, find their explanation and reason of being. Either they have done service in the past, or they may do service in the future. They may have been essentially useful in one way in a past species, and, though now functionless, they may be turned to useful account in some very different way hereafter. In botany several cases come to our mind which suggest such interpretation.

Under this view, moreover, waste of life and material in organic

Nature ceases to be utterly inexplicable, because it ceases to be objectless. It is seen to be a part of the general "economy of Nature," a phrase which has a real meaning. One good illustration of it is furnished by the pollen of flowers. The seeming waste of this in a pine-forest is enormous. It gives rise to the so-called "showers of sulphur," which everyone has heard of. Myriads upon myriads of pollen-grains (each an elaborate organic structure) are wastefully dispersed by the winds to one which reaches a female flower and fertilizes a seed. Contrast this with one of the close-fertilized flowers of a violet, in which there are not many times more grains of pollen produced than there are of seeds to be fertilized; or with an orchis-flower, in which the proportion is not widely different. These latter are certainly the more economical; but there is reason to believe that the former arrangement is not wasteful. . . . The greater economy in orchis-flowers is accounted for by the fact that the pollen is packed in coherent masses, all attached to a common stalk, the end of which is expanded into a sort of button, with a glutinous adhesive face (like a bit of sticking-plaster), and this is placed exactly where the head of a moth or butterfly will be pressed against it when it sucks nectar from the flower, and so the pollen will be bodily conveyed from blossom to blossom, with small chance of waste or loss. The floral world is full of such contrivances; and while they exist the doctrine of purpose or final cause is not likely to die out. Now, in the contrasted case, that of pine-trees, the vast superabundance of pollen would be sheer waste if the intention was to fertilize the seeds of the same tree, or if there were any provision for insect-carriage; but with wide-breeding as the end, and the wind which "bloweth where it listeth" as the means, no one is entitled to declare that pine-pollen is in wasteful excess. The cheapness of wind-carriage may be set against the over-production of pollen.

Darwinian teleology has the special advantage of accounting for the imperfections and failures as well as for successes. It not only accounts for them, but turns them to practical account. It explains the seeming waste as being part and parcel of a great economical process. Without the competing multitude, no struggle for life; and without this, no natural selection and survival of the fittest, no continuous adaptation to changing surroundings, no diversification and improvement, leading from lower up to higher and nobler forms. So the most puzzling things of all to the old-school teleologists are the *principia* of the Darwinian. In this system the forms and species, in all their variety, are not mere ends in themselves, but the whole a series of means and ends, in the contemplation of which we may obtain higher and more comprehensive, and perhaps worthier, as well as more consistent, views of design in Nature than heretofore. At least, it would appear that in Darwinian evolution we may have a theory that accords with if it does not explain the principal facts, and a teleology that is free from the common objections.

* * *

Natural selection is not the wind which propels the vessel, but the rudder which, by friction, now on this side and now on that, shapes the course. The rudder acts while the vessel is in motion, effects nothing when it is at rest. Variation answers to the wind: "Thou hearest the sound thereof, but canst not tell whence it cometh and whither it goeth." Its course is controlled by natural selection, the action of which, at any given moment, is seemingly small or insensible; but the ultimate results are great. This proceeds mainly through outward influences. But we are more and more convinced that variation, and therefore the ground of adaptation, is not a product of, but a response to, the action of the environment. Variations, in other words, the differences between individual plants and animals, however originated, are evidently not from without but from within—not physical but physiological.

The origination, and even the variation, still remains unexplained either by the action of insects or by any of the processes which collectively are personified by the term natural selection. We really believe that these exquisite adaptations have come to pass in the course of Nature, and under natural selection, but not that natural selection alone explains or in a just sense originates them. Or rather, if this term is to stand for sufficient cause and rational explanation, it must denote or include that inscrutable something which produces —as well as that which results in the survival of—"the fittest."

We have been considering this class of questions only as a naturalist might who sought for the proper or reasonable interpretation of the problem before him, unmingled with considerations from any other source. Weightier arguments in the last resort, drawn from the intellectual and moral constitution of man, lie on a higher plane, to which it was unnecessary for our particular purpose to rise, however indispensable this be to a full presentation of the evidence of mind in Nature. To us the evidence, judged as impartially as we are capable of judging, appears convincing. But, whatever view one unconvinced may take, it cannot remain doubtful what position a theist ought to occupy. If he cannot recognize design in Nature because of evolution, he may be ranked with those of whom it was said, "Except ye see signs and wonders ye will not believe." How strange that a convinced theist should be so prone to associate design only with miracle!

Baden Powell
THE VALIDATION OF RELIGION APART FROM RATIONAL PROOF

This reading and the next reflect how religious intellectuals reshaped their faith along lines compatible with modern learning and Darwinian science. The reading here, from an essay by Baden Powell (1796–1860), first appeared in Essays and Reviews *(1860), published almost simultaneously with Darwin's* Origin of Species. Essays and Reviews *contained progressive, exceedingly controversial essays by a group of liberal Anglican clergymen and represented the beginning of a new era in adjusting Christianity to modern science, biblical criticism, and historical study. As apparent in the reading, although he regarded Darwin's work as a dramatic confirmation of his point of view, Powell drew upon the* Origin *as additional support for an outlook that had been in the making for some time. Powell's essay was directed against the reigning presupposition at the time regarding the validation of religious beliefs by a series of "external" proofs (or evidences): for example, the "proof" that biblical miracles had occurred because they were reported by numerous, trustworthy, objective witnesses. This "proof" was argued classically in William Paley's famous* Evidences for Christianity *(1794)—not to be confused with his* Natural Theology, *referred to above.*

Powell was convinced that this and other external proofs, along with all attempts to "harmonize" the Bible and science, were defunct. With Samuel Taylor Coleridge (1722–1834), he believed that religion was "true" on "internal" or intrinsic grounds, like its moral, aesthetic, and emotional meaningfulness—or in Powell's words, on "spiritual" bases apart from "physical things." As a professor of geometry at Oxford University, Powell was a prominent scientist as well as an innovative theologian.

The investigation of that important and extensive subject which includes what have been usually designated as "The Evidences of Revelation," has prescriptively occupied a considerable space in the field of theological literature, especially as cultivated in England. There is scarcely one, perhaps, of our more eminent divines, who has not, in a greater or less degree, distinguished himself in this department; and scarcely an aspirant for theological distinction who has not thought it one of the surest paths to that eminence, combining so many and varied motives of ambition, to come forward as a champion in this arena. At the present day, it might be supposed the

Abridged from the American edition of *Essays and Reviews,* ed. by Frederic H. Hedge (Boston: Walker, Wise, 1861).

discussion of such a subject, taken up as it has been successively in all its conceivable different bearings, must be nearly exhausted. It must, however, be borne in mind that . . . these *external accessories* constitute a subject which of necessity is perpetually taking somewhat, at least, of a new form with the successive phases of opinion and knowledge. And it thus becomes not an unsatisfactory nor unimportant object, from time to time, to review the condition in which the discussion stands, and to comment on the peculiar features which at any particular epoch it most prominently presents, as indicative of strength or weakness—of the advance and security of the cause—if, in accordance with the real progress of enlightenment, its advocates have had the wisdom to rescind what better information showed defective, and to substitute views in accordance with higher knowledge; or, on the other hand, inevitable symptoms of weakness and inefficiency, if such salutary cautions have been neglected. To offer some general remarks of this kind on the existing state of these discussions will be the object of the present essay.

The present discussion is not intended to be of a controversial kind: it is purely contemplative and theoretical. It is rather directed to a calm and unprejudiced survey of the various opinions and arguments adduced, whatever may be their ulterior tendency, on these important questions; and to the attempt to state, analyze, and estimate them, just as they may seem really conducive to the high object professedly in view.

* * *

The scope and character of the various discussions raised on "the evidences of religion" have varied much in different ages; following, of course, both the view adopted of revelation itself, the nature of the objections which for the time seemed most prominent, or most necessary to be combated, and stamped with the peculiar intellectual character and reasoning tone of the age to which they belonged.

The early apologists were rather defenders of the Christian cause generally; but, when they entered on evidential topics, naturally did so rather in accordance with the prevalent modes of thought, than with what would now be deemed a philosophic investigation of alleged facts and critical appreciation of testimony in support of them.

The Validation of Religion Apart from Rational Proof   125

In subsequent ages, as the increasing claims of infallible church authority gained ground, to discuss evidence became superfluous, and even dangerous and impious. Accordingly, of this branch of theological literature (unless in the most entire subjection to ecclesiastical dictation) the medieval church presented hardly any specimens.

It was not perhaps till the fifteenth century that any works, bearing the character of what are now called treatises on "the evidences," appeared; and these were probably elicited by the sceptical spirit which had already begun to show itself, arising out of the subtleties of the schoolmen.

But in modern times, and under Protestant auspices, a greater disposition to follow up this kind of discussion has naturally been developed. The sterner genius of Protestantism required definition, argument, and proof, where the ancient church had been content to impress by the claims of authority, veneration, and prescription, and thus left the conception of truth to take the form of a mere impression of devotional feeling or exalted imagination.

Protestantism sought something more definite and substantial; and its demands were seconded and supported, more especially by the spirit of metaphysical reasoning which so widely extended itself in the seventeenth century, even into the domains of theology; and divines, stirred up by the allegations of the Deists, aimed at formal refutations of their objections, by drawing out the idea and the proofs of revelation into systematic propositions supported by logical arguments. In that and the subsequent period, the same general style of argument on these topics prevailed among the advocates of the Christian cause. The appeal was mainly to the miracles of the Gospels; and here, it was contended, we want merely the same testimony of eyewitnesses which would suffice to substantiate any ordinary matter of fact. Accordingly, the narratives were to be traced to writers at the time, who were either themselves eyewitnesses, or recorded the testimony of those who were so; and, the direct transmission of the evidence being thus established, everything was held to be demonstrated. If any antecedent question was raised, a brief reference to the Divine Omnipotence to work the miracles, and to the Divine Goodness to vouchsafe the revelation and confirm it by such proofs, was all that could be required to silence skeptical cavils.

It is true, indeed, that some consideration of the *internal* evidence derived from the excellence of the doctrines and morality of the gospel was allowed to enter the discussion; but it formed only a subordinate branch of the evidences of Christianity. The main and essential point was always the consideration of external facts, and the attestations of testimony offered in support of them. Assuming Christianity to be essentially connected with certain outward and sensible events, the main thing to be inquired into and established was the historical evidence of those events, and the genuineness of the records of them. If this were satisfactorily made out, then it was considered the object was accomplished.

The difficulties with respect to miraculous evidence in particular will necessarily be very differently viewed in different stages of philosophical and physical information. Difficulties in the idea of suspensions of natural laws, in former ages were not at all felt, canvassed, or thought of; but, in later times, they have assumed a much deeper importance. In an earlier period of our theological literature, the critical investigation of the question of *miracles* was a point scarcely at all appreciated. The attacks of the Deists of the seventeenth and early part of the eighteenth century were almost wholly directed to other points: but the speculations of Woolston, and, still more, the subsequent influence of the celebrated Essay of Hume, had the effect of directing the attention of divines more pointedly to the precise topic of miraculous evidence; and to these causes was added the agitation of the question of the ecclesiastical miracles.

In appreciating the evidence for *any* events of a striking or wonderful kind, we must bear in mind the extreme difficulty which always occurs in eliciting the truth, dependent not on the uncertainty in the transmission of testimony, but, even in cases where we were ourselves witnesses, on the enormous influence exerted by our prepossessions previous to the event, and by the momentary impressions consequent upon it. We look at all events through the medium of our prejudices; or, even where we may have no prepossessions, the more sudden and remarkable any occurrence may be, the more unprepared we are to judge of it accurately or to view it calmly. Our after-representations, especially of any extraordinary and striking event, are always, at the best, mere recollections of our impressions, of ideas dictated by our emotions at the time, of surprise and astonishment which the suddenness and hurry of the occurrence did not

The Validation of Religion Apart from Rational Proof

allow us time to reduce to reason, or to correct by the sober standard of experience or philosophy.

We can only hope to form just and legitimate conclusions from an extended and unprejudiced study of the laws and phenomena of the natural world. The entire range of the inductive philosophy is at once based upon, and in every instance tends to confirm by immense accumulation of evidence, the grand truth of the universal order and constancy of natural causes as a primary law of belief; so strongly entertained and fixed in the mind of every truly inductive inquirer, that he can hardly even conceive the possibility of its failure. Yet we sometimes hear language of a different kind. There are still some who dwell on the idea of Spinoza, and contend that it is idle to object to miracles as violations of natural laws, because we know not the extent of nature; that all inexplicable phenomena are, in fact, miracles, or, at any rate, mysteries; that we are surrounded by miracles in nature, and on all sides encounter phenomena which baffle our attempts at explanation, and limit the powers of scientific investigation—phenomena whose causes or nature we are not, and probably never shall be, able to explain.

Such are the arguments of those who have failed to grasp the positive scientific idea of the power of the inductive philosophy, or the *order of nature*. The boundaries of nature exist only where our *present* knowledge places them: the discoveries of tomorrow will alter and enlarge them. The inevitable progress of research must, within a longer or shorter period, unravel all that seems most marvelous; and what is at present least understood will become as familiarly known to the science of the future, as those points which a few centuries ago were involved in equal obscurity, but are now thoroughly understood.

The enlarged critical and inductive study of the natural world cannot but tend powerfully to evince the inconceivableness of imagined interruptions of natural order or supposed suspensions of the laws of matter, and of that vast series of dependent causation which constitutes the legitimate field for the investigation of science, whose constancy is the sole warrant for its generalizations.

By those who take a more enlarged survey of the subject, it cannot fail to be remarked how different has been the spirit in which miracles were contemplated, as they are exhibited to us in the earlier stages of ecclesiastical literature, from that in which they have been regarded in modern times; and this especially in respect to that particular view which has so intimately connected them with precise "evidential arguments," and by a school of writers, of whom Paley may be taken as the type, and who regard them as the sole external proof and certificate of a divine revelation.

But, at the present day, this "evidential" view of miracles as the sole, or even the principal, external attestation to the claims of a divine revelation, is a species of reasoning which appears to have lost ground even among the most earnest advocates of Christianity. It is now generally admitted that Paley took too exclusive a view in asserting that we cannot conceive a revelation substantiated in any other way; and it has been even more directly asserted by some zealous supporters of Christian doctrine, that the external evidences are altogether inappropriate and worthless.

The poet Coleridge—than whom no writer has been more earnest in upholding and defending Christianity, even in its most orthodox form—in speaking of its external attestations, impatiently exclaims, "Evidences of Christianity! I am weary of the word. Make a man feel the want of it, . . . and you may safely trust it to its own evidence."

A considerable school have been disposed to look to the intrinsic evidence *only,* and to accept the declarations of the gospel *solely* on the ground of their intrinsic excellence, and *accordance* with our best and highest moral and religious convictions; a view which would approach very nearly to rejecting its peculiarities altogether.

Those who have reflected most deeply on the nature of the argument from external evidence will admit, that it would naturally possess very different degrees of force as addressed to different ages; and, in a period of advanced physical knowledge, the reference to what was believed in past times, if at variance with principles now acknowledged, could afford little ground of appeal; in fact, would damage the argument rather than assist it.

* * *

All reason and science conspire to the confession, that, beyond the domain of physical causation and the possible conceptions of *intellect* or *knowledge,* there lies open the boundless region of spiritual things, which is the sole dominion of *faith;* and while intellect

The Validation of Religion Apart from Rational Proof

and philosophy are compelled to disown the recognition of anything in the world of matter at variance with the first principle of the laws of matter—the universal order and indissoluble unity of physical causes—they are the more ready to admit the higher claims of divine mysteries in the invisible and spiritual world. Advancing knowledge, while it asserts the dominion of science in physical things, confirms that of faith in spiritual: we thus neither impugn the generalizations of philosophy, nor allow them to invade the dominion of faith, and admit that what is not a subject for a problem may hold its place in a creed.

In an evidential point of view, it has been admitted by some of the most candid divines, that the appeal to miracles, however important in the early stages of the gospel, has become less material in later times; and others have even expressly pointed to this as the reason why they have been withdrawn: whilst, at the present day, the most earnest advocates of evangelical faith admit that outward marvels are needless to spiritual conviction, and triumph in the greater moral miracle of a converted and regenerate soul.

They echo the declaration of St. Chrysostom, "If you are a believer as you ought to be, and love Christ as you ought to love him, you have no need of miracles; for these are given to unbelievers."

After all, the *evidential* argument has but little actual weight with the generality of believers.

Matters of clear and positive fact, investigated on critical grounds and supported by exact evidence, are properly matters of knowledge, not of faith. It is rather in points of less definite character that any exercise of faith can take place; it is rather with matters of religious belief, belonging to a higher and less conceivable class of truths, with the mysterious things of the unseen world, that faith owns a connection, and more readily associates itself with spiritual ideas, than with external evidence or physical events: and it is generally admitted, that many points of important religious instruction, even conveyed under the form of fictions—as in the instances of doctrines inculcated through parables—are more congenial to the spirit of faith than any relations of historical events could be.

The more knowledge advances, the more it has been, and will be, acknowledged that Christianity, as a real religion, must be viewed apart from connection with physical things.

The first dissociation of the spiritual from the physical was rendered necessary by the palpable contradictions disclosed by astronomical discovery with the letter of Scripture. Another still wider and more material step has been effected by the discoveries of geology. More recently, the antiquity of the human race and the development of species, and the rejection of the idea of "creation," have caused new advances in the same direction.

In all these cases, there is, indeed a direct discrepancy between what had been taken for revealed truth and certain undeniable existing monuments to the contrary.

But these monuments were interpreted by science and reason; and there are other deductions of science and reason referring to alleged events, which, though they have left no monuments or permanent effects behind them, are not the less legitimately subject to the conclusions of positive science, and require a similar concession and recognition of the same principle of the independence of spiritual and of physical truth.

Thus far, our observations are general; but, at the present moment, some recent publications on the subject seem to call for a few more detailed remarks. We have before observed, that the style and character of works on "the evidences" has, of necessity, varied in different ages. Those of Leslie and Grotius have, by common consent, been long since superseded by that of Paley. Paley was long the textbook at Cambridge: his work was never so extensively popular at Oxford; it has of late been entirely disused there. By the public at large however once accepted, we do not hesitate to express our belief, that, before another quarter of a century has elapsed, it will be laid on the shelf with its predecessors: not that it is a work destitute of high merit—as is preeminently true also of those it superseded, and of others again anterior to them—but they have all followed the irreversible destiny, that a work, suited to convince the public mind at any one particular period, must be accommodated to the actual condition of knowledge, of opinion, and mode of thought, of that period. It is not a question of *abstract excellence,* but of *relative adaptation.*

Paley caught the prevalent tone of thought in his day. Public opinion has now taken a different turn; and, what is more important, the style and class of difficulties and objections *honestly* felt has become wholly different. New modes of speculation—new forms of skepticism—have invaded the domain of that settled belief which a

The Validation of Religion Apart from Rational Proof

past age had been accustomed to rest on the Paleyean syllogism. Yet, among several works which have of late appeared on the subject, we recognize few which at all meet these requirements of existing opinion. Of some of the chief of these works, even appearing under the sanction of eminent names, we are constrained to remark, that they are altogether behind the age. . . .

In truth, the majority of these champions of the evidential logic betray an almost entire unconsciousness of the advance of opinion around them. Having their own ideas long since cast in the stereotyped mould of the past, they seem to expect that a progressing age ought still to adhere to the same type, and bow implicitly to a solemn and pompous but childish parade and reiteration of the one-sided dogmas of an obsolete school, coupled with awful denunciations of heterodoxy on all who refuse to listen to them.

* * *

In an age of physical research like the present, all highly cultivated minds and duly advanced intellects have imbibed, more or less, the lessons of the inductive philosophy, and have, at least in some measure, learned to appreciate the grand-foundation conception of universal law; to recognize the impossibility . . . of any modification whatsoever in the existing conditions of material agents, unless through the invariable operation of a series of eternally impressed consequences, following in some necessary chain of orderly connection, however imperfectly known to us.

A work has now appeared by a naturalist of the most acknowledged authority—Mr. Darwin's masterly volume on *The Origin of Species* by the law of "natural selection"—which now substantiates on undeniable grounds the very principle so long denounced by the first naturalists—*the origination of new species by natural causes;* a work which must soon bring about an entire revolution of opinion in favor of the grand principle of the self-evolving powers of nature.

The main assertion of Paley is, that it is impossible to conceive a revelation given except by means of miracles. This is his primary axiom; but this is precisely the point which the modern turn of reasoning most calls in question, and rather adopts the belief that a revelation is then most credible, when it appeals least to violations of natural causes. Thus, if miracles were, in the estimation of a former age, among the chief *supports* of Christianity, they are at present among the main *difficulties,* and hindrances to its acceptance.

In nature and from nature, by science and by reason, we neither have, nor can possibly have, any evidence of a *Deity working miracles:* for that, we must go out of nature and beyond reason. If we could have any such evidence *from nature,* it could only prove extraordinary *natural* effects, which would not be *miracles* in the old theological sense, as isolated, unrelated, and uncaused; whereas no physical fact can be conceived as unique, or without analogy and relation to others and to the whole system of natural causes.

The "*reason* of the hope that is in us" is not restricted to *external* signs, nor to any one kind of evidence, but consists of such assurance as may be most satisfactory to each earnest individual inquirer's own mind: and the true acceptance of the entire revealed mainfestation of Christianity will be most worthily and satisfactorily based on that assurance of "faith," by which, the apostle affirms, "we stand" (2 Cor. ii. 24); and which, in accordance with his emphatic declaration, must rest, "not in the wisdom of man, but in the power of God" (1 Cor. ii. 5).

S. R. Driver
A NEW UNDERSTANDING OF THE BIBLE

As a liberal biblical scholar with an international reputation, S. R. Driver (1846–1914) spent a lifetime seeking to convince the English-speaking world to accept the conclusions of radical biblical criticism. That criticism entailed a reconstruction of traditional opinions about the authorship, literary composition, and historical accuracy of the Bible, and added its own set of dramatic challenges to traditional religion.

One of the reasons why Driver sought to popularize biblical criticism was based on his personal belief that it did not undermine religious faith. In the selection here, from an article on the cosmology of Genesis, Driver argues that the biblical creation stories reflect the creation narratives of the ancient Near East and do "not accord with the results established by science." Yet even though Driver thought that attempts to harmonize Genesis and science were invalidated, he sought to establish a new kind of compatibility between the two by saying that the purposes of the writers of the Bible were purely religious and that the six days of creation were an intentionally idealized, "representative" depiction of the creation.

In this sense biblical criticism, although radical by traditional standards, appeared to offer a response to the charge that geology and evolution had proved that the Bible was false. At the hands of Driver and other biblical critics, the Bible was credited with nonliteral, nonscientific intentionality and integrity. How convincing is this point of view?

Are we any nearer than we were to a reconciliation of Genesis and science? And, if not, what position is the theologian to assume, and in what light is he to view the familiar and impressive narrative with which the Bible opens?

The past history of our earth is known approximately by evidence which cannot be gainsaid—the evidence engraven in the rocks. Those cliffs which tower out of the sea on our southern coasts have revealed to the microscope the secret of their growth: they are composed of the minute shells of marine organisms, deposited at the rate of a few inches a century at the bottom of the ocean, and afterwards, by some great upheaval of the earth's crust, lifted high above the waves. Our coal measures are the remains of mighty forests which, one after another in slow succession, have come and gone in certain parts of the earth's surface, and have stored up the

Abridged from S. R. Driver, "The Cosmogony of Genesis," *The Expositor* 3 (January 1886).

energy, poured forth during long ages from the sun, for our consumption and enjoyment.

Geology has become a science; and the indications which have been noticed, with countless others, show that the earth was not created, substantially as we know it, some 6000 years ago, but that it reached its present state, and received its rich and wondrous adornment of vegetable and animal life, by a gradual process, extending over untold centuries, and embracing unnumbered generations of living forms. More than this, not only do geology and paleontology trace the history of the earth's crust, and determine the succession of living forms which have peopled it, but astronomy, comparing the system of which this globe forms part with other systems, takes a bolder flight, and rises to the conception of a theory explaining, by the aid of known mechanical and physical principles, the formation of the earth itself. Observing the nature of the sun and of the planets, and other countless small bodies revolving round it; perceiving, by the spectroscope and other means, that the elements of which all are composed are similar, and assured by the nebulae of the existence in the heavens of huge masses of luminous gas; astronomers following Laplace have supposed that the substance of which the solar system is composed existed once as a diffused gaseous mass, which gradually condensed and became a rotating sphere, from which, in succession, the different planets were flung off, while the remainder was more and more concentrated until it became what we call the sun. One of these planets, our earth—we need affirm nothing respecting the others—in course of time, by reduction of temperature, and otherwise, developed the conditions adequate for the support of life.

. . . Let us proceed at once to compare the process by which, according to the narrative in Genesis, the earth was fitted to become the habitation of man, with that which is disclosed by the investigations of science. In the first place, since the fossil remains embedded in the different strata of the earth's surface show, beyond reach of controversy, that the living forms which preceded man upon this globe were distributed in a definite order over periods of vast duration, we must, if we suppose this order to be described in Genesis, inquire whether it is permissible to understand the term *day* in any but its literal sense. *In the representation of the writer* it seems clear

A New Understanding of the Bible

that the term denotes a period of twenty-four hours. The passages which have been adduced to establish the contrary are inconclusive.

If . . . at least provisionally, *day* be interpreted as equivalent to *period,* two questions at once arise: Do the days of Genesis correspond with well-defined geological periods? and does the order in which different living things are stated to have been created agree with the facts of geology? To both these questions candor compels the answer, No.

(1) The geological record contains no evidence of clearly defined periods, corresponding to the days of Genesis. This, however, may be considered a minor discrepancy. (2) In Genesis vegetation is complete two days before animal life appears: geology shows that they appear simultaneously—even if animal life does not appear first. (3) In Genesis birds appear together with aquatic creatures, and precede all land animals: according to the evidence of geology, birds are unknown till a period much later than that at which aquatic creatures (including fishes and amphibia) abound, and they are preceded by numerous species of land animals—in particular by insects, and other "creeping things." The second and third of these discrepancies are formidable.

However reluctant we may be to make the admission, only one conclusion seems possible. Read without prejudice or bias, the narrative of Genesis i. creates an impression *at variance with the facts revealed by science:* efforts of reconciliation which have been reviewed are different modes of obliterating its characteristic features, and of reading into it a *view which it does not express.*

What then may we suppose to be the source of the cosmogony in Genesis? In answering this question we must bear in mind the position which the Hebrews took among the nations of antiquity. In the possession of aptitudes fitting them in a peculiar measure to become the organ and channel of revelation, the Hebrew nation differed radically from its neighbors; but it was allied to them in language, it shared with them many of the same institutions, the same ideas and habits of thought. Other nations of antiquity made efforts to fill the void in the past which begins where historical reminiscences cease; and framed theories to account for the beginnings of the earth and man, or to solve the problems which the observation of human nature suggests. It is but consonant with analogy to suppose that the Hebrews either did the same for themselves, or borrowed those of their neighbors. Of the theories current in Assyria and Phoenicia, fragments have been preserved, and these exhibit points of resemblance with the Biblical narrative sufficient to warrant the inference that both are derived from the same cycle of tradition. Here are three fragments from the "Creation Tablets," belonging to the library of Asshurbanipal (668–626 B.C.), discovered by the late George Smith:

> *When as yet the heavens above had not declared,*
> *Nor the earth beneath had recorded a name,*
> *The august ocean was their generator,*
> *The surging deep was she that bare them all,*
> *The waters thereof embraced one another and united,*
> *But darkness was not yet withdrawn, nor had vegetation sprung forth.*
>
> *When of the gods none yet had issued forth,*
> *Or recorded a name, or [fixed] a destiny,*
> *Then were the [great] gods formed.*
> *The gods Lachmu and Lachamu proceeded forth.*
> . . .
> *He made beautiful the dwellings of the great gods.*
> *The stars, likewise, he caused . . . come forth:*
> *He ordained the year, established for it decades,*
> *Brought forth the twelve months each with three stars.*
> . . .
> *When the gods in their assembly formed . . .*
> *They made beautiful the mighty [trees?],*
> *And caused living beings to come forth. . . .*

From a theological point of view, this is different enough from the Biblical record; at the same time, side by side with the difference, there are material resemblances which cannot be mistaken. We have, for instance, the same idea of a surging chaos, reduced gradually to order, the same view of the appointment of years and seasons, and of the formation subsequently of living creatures. Similarly, the Phoenician traditions, which were translated into Greek by Philo of Byblus, and are preserved to us in their Greek form by Eusebius, describe the origin of different institutions and inventions, in a style which at once recalls that of the latter part of the fourth chapter of Genesis. In the light of these facts it is difficult to resist the conclusion that the Biblical narrative is drawn from

A New Understanding of the Bible

the same source as these other records. . . . The discrepancies that have been dwelt upon—and which, so far as can be seen, appear irremovable—seem to constitute an indication that the cosmogony of Genesis is not meant to be an authoritative exposition of the past history of the earth, but that it subserves a different purpose altogether. Its purpose is to teach *religious* truth, not scientific truth. With this object in view, its author sets before us a series of *representative pictures,* remarkably suggestive of the reality, if only they be not treated as a "revelation" of it, and embodying theological teaching of permanent value. It only remains to indicate briefly the nature of this teaching.

1. It shows in opposition to the conceptions prevalent in antiquity, that the world is not self-originated; that it was called into existence, and brought gradually into its present state, at the will of a spiritual Being, prior to it, independent of it, and deliberately planning every stage of its progress. It is this feature which distinguishes it fundamentally from the Babylonian cosmogony, with which, as we have seen, it bears an external resemblance. The Babylonian scheme is essentially polytheistic; chaos is anterior to Deity; the gods are made, or produced—we know not whence or how. In Genesis, the supremacy of the Creator is absolute.
2. Dividing artificially the entire period into six parts, it notices in order the most prominent cosmical phenomena, and groups the living creatures upon the earth under the great subdivisions which appeal to the eye. By this method it exhibits an *ideal picture* of the successive stages by which the earth was formed and peopled with its living inhabitants; and it insists that each of these stages is no product of chance, or of mere mechanical forces, but is an act of the Divine will, realizes the Divine purpose, and receives the seal of the Divine approval. The slow formation of the earth, as taught by geology, the gradual development of species made probable by modern biology, is but the exhibition in detail of those processes which the author of this cosmogony sums up into a single phrase and apparently compresses into a single moment, for the purpose of declaring their dependence on the Divine will.
3. It insists on the distinctive preeminence belonging to man, implied in the remarkable self-deliberation taken in his case by tne Creator, and signified expressly in the phrase, "image

of God," by which doubtless is meant the possession by man of *self-conscious reason*—an adumbration, we may suppose, however faint, of the supreme mind of God—enabling him to *know,* in a sense in which animals do not know, and involving the capacity of apprehending moral and religious truth.

The efforts of the harmonists have been praiseworthy, and well-meaning, but they have resulted only in the construction of artificial schemes, the unreality of which is at once detected by the scientific mind, and creates a prejudice against the entire system with which the cosmogony is connected. The cosmogony of Genesis is treated in popular estimation as an integral element of the Christian faith. It cannot be too earnestly represented that this is not the case. A definition of the process by which, after it was created, the world assumed its present condition, forms no element in the Christian creed. The Church has never pronounced with authority upon the interpretation of the narrative of Genesis. It is our duty to eradicate popular illusions, and to teach *both* that the cosmogony of Genesis does not accord with the results established by science, *and* that the recognition of this fact is no invasion of sacred ground, and in no degree imperils the Christian revelation. . . . [The Biblical narrative] neither comes into collision with science, nor needs reconciliation with it; its office lies in a different plane altogether; it is to present, under a form impressive to the imagination, adapted to the needs of all time, and containing no feature unworthy of the dignity of its subject, a truthful *representative picture* of the relation of the world to God.

Thomas Hardy and Alfred Lord Tennyson
THE NEW FACES OF NATURE

These poems by Thomas Hardy (1840–1928) and Alfred Lord Tennyson (1809–1892) reflect the vast emotional and intellectual revolution set in motion by evolution and science. In their perceptions of nature, their lack of a sense of purpose and peace, their uneasiness in trusting God or traditional values, Hardy and Tennyson lie poles removed from William Paley, William Cowper, and William Wordsworth. Finding themselves surrounded by a nature "red in tooth and claw," Hardy and Tennyson neither derived comfort nor received moral instruction from tranquil, wooded landscapes or a "sportive train" of wild animals. Hardy lost his romantic faith upon reading Darwin's Origin of Species *shortly after his twentieth birthday and could no longer regard nature as the product of a Creator who cared for man and his values. Tennyson wrote* In Memoriam, *from which the selections here are taken, almost a decade before the* Origin, *but the poem reflected so uncannily the problems introduced and encouraged by Darwinism, that it became immensely meaningful to generations of thinkers who were forced to adjust their ideas and feelings to the new age. With a chastened, abridged faith and surrounded with doubts, Tennyson sought for purpose and meaning in a moral nature residing in mankind and in a mystical communion with immortal souls like his friend Arthur Henry Hallam, for whom he wrote* In Memoriam.

In a Wood

Pale beech and pine so blue,
   Set in one clay,
Bough to bough cannot you
   Live out your day?
When the rains skim and skip,
Why mar sweet comradeship,
Blighting with poison-drip
   Neighbourly spray?

Heart-halt and spirit-lame,
   City-opprest,
Unto this wood I came
   As to a nest;

"In a Wood" and "The Darkling Thrush" are from Thomas Hardy, *Collected Poems* (New York: Macmillan, 1926). Printed by the permission of the Trustees of the Hardy Estate, The Macmillan Company of Canada, Limited; and Macmillan, London and Basingstoke.

"In Memoriam A. H. H." is from Alfred Lord Tennyson, *The Poetic and Dramatic Works of Alfred Lord Tennyson* (Boston: Houghton Mifflin, 1898).

Dreaming that sylvan peace
Offered the harrowed ease—
Nature a soft release
   From men's unrest.

But, having entered in,
   Great growths and small
Show them to men akin—
   Combatants all!
Sycamore shoulders oak,
Bines the slim sapling yoke,
Ivy-spun halters choke
   Elms stout and tall.

Touches from ash, O wych,
   Sting you like scorn!
You, too, brave hollies, twitch
   Sidelong from thorn.
Even the rank poplars bear
Lothly a rival's air,
Cankering in black despair
   If overborne.

Since, then, no grace I find
   Taught me of trees,
Turn I back to my kind,
   Worthy as these.
There at least smiles abound,
There discourse trills around,
There, now and then, are found
   Life-loyalties.

The Darkling Thrush

I leant upon a coppice gate
   When Frost was spectre-gray,
And Winter's dregs made desolate
   The weakening eye of day.
The tangled bine-stems scored the sky
   Like strings of broken lyres,
And all mankind that haunted nigh
   Had sought their household fires.

The land's sharp features seemed to be
   The Century's corpse outleant,
His crypt the cloudy canopy,
   The wind his death-lament.

The ancient pulse of germ and birth
    Was shrunken hard and dry,
And every spirit upon earth
    Seemed fervourless as I.

At once a voice arose among
    The bleak twigs overhead
In a full-hearted evensong
    Of joy illimited;
An aged thrush, frail, gaunt, and small,
    In blast-beruffled plume,
Had chosen thus to fling his soul
    Upon the growing gloom.

So little cause for carolings
    Of such ecstatic sound
Was written on terrestrial things
    Afar or nigh around,
That I could think there trembled through
    His happy good-night air
Some blessed Hope, whereof he knew
    And I was unaware.

In Memoriam A. H. H.
[54]

O, yet we trust that somehow good
    Will be the final goal of ill,
    To pangs of nature, sins of will,
Defects of doubt, and taints of blood;

That nothing walks with aimless feet;
    That not one life shall be destroy'd,
    Or cast as rubbish to the void,
When God hath made the pile complete;

That not a worm is cloven in vain;
    That not a moth with vain desire
    Is shrivell'd in a fruitless fire,
Or but subserves another's gain.

Behold, we know not anything;
    I can but trust that good shall fall
    At last—far off—at last, to all,
And every winter change to spring.

So runs my dream; but what am I?
    An infant crying in the night;
    An infant crying for the light,
And with no language but a cry.

[55]

The wish, that of the living whole
    No life may fail beyond the grave,
    Derives it not from what we have
The likest God within the soul?

Are God and Nature then at strife,
    That Nature lends such evil dreams?
    So careful of the type she seems,
So careless of the single life,

That I, considering everywhere
    Her secret meaning in her deeds,
    And finding that of fifty seeds
She often brings but one to bear,

I falter where I firmly trod,
    And falling with my weight of cares
    Upon the great world's altar-stairs
That slope thro' darkness up to God,

I stretch lame hands of faith, and grope,
    And gather dust and chaff, and call
    To what I feel is Lord of all,
And faintly trust the larger hope.

[56]

"So careful of the type?" but no.
    From scarped cliff and quarried stone
    She cries, "A thousand types are gone;
I care for nothing, all shall go.

"Thou makest thine appeal to me.
    I bring to life, I bring to death;
    The spirit does but mean the breath:
I know no more." And he, shall he,

Man, her last work, who seem'd so fair,
    Such splendid purpose in his eyes,
    Who roll'd the psalm to wintry skies,
Who built him fanes of fruitless prayer,

Who trusted God was love indeed
    And love Creation's final law—
    Tho' Nature, red in tooth and claw
With ravine, shriek'd against his creed—

Who loved, who suffer'd countless ills,
    Who battled for the True, the Just,
    Be blown about the desert dust,
Or seal'd within the iron hills?

No more? A monster then, a dream,
    A discord. Dragons of the prime,
    That tare each other in their slime,
Were mellow music match'd with him.

O life as futile, then, as frail!
    O for thy voice to soothe and bless!
    What hope of answer, or redress?
Behind the veil, behind the veil.

[120]

I trust I have not wasted breath:
    I think we are not wholly brain,
    Magnetic mockeries; not in vain,
Like Paul with beasts, I fought with Death;

Not only cunning casts in clay:
    Let Science prove we are, and then
    What matters Science unto men,
At least to me? I would not stay.

Let him, the wiser man who springs
    Hereafter, up from childhood shape
    His action like the greater ape,
But I was born to other things.

[124]

That which we dare invoke to bless;
    Our dearest faith; our ghastliest doubt;
    He, They, One, All; within, without;
The Power in darkness whom we guess,—

I found Him not in world or sun,
    Or eagle's wing, or insect's eye,
    Nor thro' the questions men may try,
The petty cobwebs we have spun.

A warmth within the breast would melt
    The freezing reason's colder part,
    And like a man in wrath the heart
Stood up and answer'd, "I have felt."

No, like a child in doubt and fear:
    But that blind clamor made me wise;
    Then was I as a child that cries,
But, crying, knows his father near;

And what I am beheld again
    What is, and no man understands;
    And out of darkness came the hands
That reach thro' nature, moulding men.

[From "Epilogue"]

No longer half-akin to brute,
    For all we thought and loved and did,
    And hoped, and suffer'd, is but seed
Of what in them is flower and fruit;

Whereof the man that with me trod
    This planet was a noble type
    Appearing ere the times were ripe,
That friend of mine who lives in God,

That God, which ever lives and loves,
    One God, one law, one element,
    And one far-off divine event,
To which the whole creation moves.

III EVOLUTION, MAN AND SOCIETY

FIGURE 6. In this cartoon from *Punch* magazine, Darwin, robed like a Greek philosopher, contemplates the evolutionary process: arising out of CHAOS, the worm becomes an ape, and the ape evolves into a very proper Victorian.

Charles Darwin

THE NATURAL EVOLUTION OF MAN AND MORALITY

Even though The Origin of Species *contained only one reference to man—namely, the classic understatement that via evolution, light "will be thrown on the origin of man and his history"—Darwin and his friends suspected that much of the storm of debate over evolution by natural selection would focus on the status of man. Their expectations were more than fulfilled. In attempting to understand nature (and man) apart from religious ideas, Darwin appeared to sever mankind's spiritual umbilical cord, mankind's relationship to some spiritual being or nature. In linking man intrinsically to the evolution of plant and animal species, he obliterated distinctions that had previously preserved human dignity, and implied that mankind was the product of an immoral and naturalized world.*

After the release of Origin *in 1859, Darwin waited twelve years before he published his research and opinions concerning the origin and significance of man. The excerpts here, from* The Descent of Man *(1871), summarize those ideas and reveal how Darwin conceived of man's total being as the natural product of evolution, yet remained confident that man was "the wonder and glory of the Universe." Is his evaluation of man's nature and accounting of man's origins convincing?*

Bodily Structure

It is notorious that man is constructed on the same general type or model as other mammals. All the bones in his skeleton can be compared with corresponding bones in a monkey, bat, or seal. So it is with his muscles, nerves, blood-vessels and internal viscera. The brain, the most important of all the organs, follows the same law.

The homological construction of the whole frame in the members of the same class is intelligible, if we admit their descent from a common progenitor, together with their subsequent adaptation to diversified conditions. On any other view, the similarity of pattern between the hand of a man or monkey, the foot of a horse, the flipper of a seal, the wing of a bat, etc., is utterly inexplicable. It is no scientific explanation to assert that they have all been formed on

Abridged from Charles Darwin, *The Descent of Man* (New York: D. Appleton, 1898 [second ed. in 1874]).

the same ideal plan. With respect to development, we can clearly understand, on the principle of variation supervening at a rather late embryonic period, and being inherited at a corresponding period, how it is that the embryos of wonderfully different forms should still retain, more or less perfectly, the structure of their common progenitor. No other explanation has ever been given of the marvelous fact that the embryos of a man, dog, seal, bat, reptile, etc., can at first hardly be distinguished from each other. In order to understand the existence of rudimentary organs, we have only to suppose that a former progenitor possessed the parts in question in a perfect state, and that under changed habits of life they became greatly reduced, either from simple disuse, or through the natural selection of those individuals which were least encumbered with a superfluous part, aided by the other means previously indicated.

Thus we can understand how it has come to pass that man and all other vertebrate animals have been constructed on the same general model, why they pass through the same early stages of development, and why they retain certain rudiments in common. Consequently we ought frankly to admit their community of descent; to take any other view, is to admit that our own structure, and that of all the animals around us, is a mere snare laid to entrap our judgment. This conclusion is greatly strengthened, if we look to the members of the whole animal series, and consider the evidence derived from their affinities or classification, their geographical distribution and geological succession. It is only our natural prejudice, and that arrogance which made our forefathers declare that they were descended from demigods, which leads us to demur to this conclusion. But the time will before long come, when it will be thought wonderful that naturalists, who were well acquainted with the comparative structure and development of man, and other mammals, should have believed that each was the work of a separate act of creation.

Natural Selection

Man is variable in body and mind; and . . . the variations are induced, either directly or indirectly, by the same general causes, and obey the same general laws, as with the lower animals. Man has spread widely over the face of the earth, and must have been exposed, during his incessant migration, to the most diversified conditions. The inhabitants of Tierra del Fuego, the Cape of Good Hope, and Tasmania in the one hemisphere, and of the Arctic regions in the other, must have passed through many climates, and changed their habits many times, before they reached their present homes. The early progenitors of man must also have tended, like all other animals, to have increased beyond their means of subsistence; they must, therefore, occasionally have been exposed to a struggle for existence, and consequently to the rigid law of natural selection. Beneficial variations of all kinds will thus, either occasionally or habitually, have been preserved and injurious ones eliminated. I do not refer to strongly marked deviations of structure, which occur only at long intervals of time, but to mere individual differences. We know, for instance, that the muscles of our hands and feet, which determine our powers of movement, are liable, like those of the lower animals, to incessant variability. If then the progenitors of man inhabiting any district, especially one undergoing some change in its conditions, were divided into two equal bodies, the one half which included all the individuals best adapted by their powers of movement for gaining subsistence, or for defending themselves, would on an average survive in greater numbers, and procreate more offspring than the other and less well endowed half.

Man in the rudest state in which he now exists is the most dominant animal that has ever appeared on this earth. He has spread more widely than any other highly organized form: and all others have yielded before him. He manifestly owes this immense superiority to his intellectual faculties, to his social habits, which lead him to aid and defend his fellows, and to his corporeal structure. The supreme importance of these characters has been proved by the final arbitrament of the battle for life. Through his powers of intellect, articulate language has been evolved; and on this his wonderful advancement has mainly depended. . . . He has invented and is able to use various weapons, tools, traps, etc., with which he defends himself, kills or catches prey, and otherwise obtains food. He has made rafts or canoes for fishing or crossing over to neighboring fertile islands. He has discovered the art of making fire, by which hard and stringy roots can be rendered digestible, and poisonous roots or herbs innocuous. This discovery of fire, probably the greatest ever made by man, excepting language, dates from before the dawn of history. These several inventions, by which man in the rudest state has become so preeminent, are the direct results of the development of his powers of observation, memory, curiosity, imagination, and reason. I cannot, therefore, understand how it is that Mr. Wallace maintains, that "natural selection could only have endowed the savage with a brain a little superior to that of an ape."

Although the intellectual powers and social habits of man are of paramount importance to him, we must not underrate the importance of his bodily structure.

Even to hammer with precision is no easy matter, as everyone who has tried to learn carpentry will admit. To throw a stone with as true an aim as a Fuegian in defending himself, or in killing birds, requires the most consummate perfection in the correlated action of the muscles of the hand, arm, and shoulder, and, further, a fine sense of touch. In throwing a stone or spear, and in many other actions, a man must stand firmly on his feet; and this again demands the perfect co-adaptation of numerous muscles. To chip a flint into the rudest tool, or to form a barbed spear or hook from a bone, demands the use of a perfect hand.

If it be an advantage to man to stand firmly on his feet and to have his hands and arms free, of which, from his preeminent success in the battle of life, there can be no doubt, then I can see no reason why it should not have been advantageous to the progenitors of man to have become more and more erect or bipedal. They would thus have been better able to defend themselves with stones or clubs, to attack their prey, or otherwise to obtain food. The best built individuals would in the long run have succeeded best, and have survived in larger numbers.

In regard to bodily size or strength, we do not know whether man is descended from some small species, like the chimpanzee, or from one as powerful as the gorilla; and, therefore, we cannot say whether man has become larger and stronger, or smaller and weaker, than his ancestors. We should, however, bear in mind that an animal possessing great size, strength, and ferocity, and which, like the gorilla, could defend itself from all enemies, would not perhaps have become social: and this would most effectually have checked the acquirement of the higher mental qualities, such as sympathy and the love of his fellows. Hence it might have been an immense advantage to man to have sprung from some comparatively weak creature.

Mind, Emotions, Religion

We have seen . . . that man bears in his bodily structure clear traces of his descent from some lower form; but it may be urged that, as man differs so greatly in his mental power from all other animals, there must be some error in this conclusion. No doubt the difference in this respect is enormous, even if we compare the mind of one of the lowest savages, who has no words to express any number higher than four, and who uses hardly any abstract terms for common objects or for the affections, with that of the most highly organized ape. The difference would, no doubt, still remain immense, even if one of the higher apes had been improved or civilized as much as a dog has been in comparison with its parent-form, the wolf or jackal. The Fuegians rank amongst the lowest barbarians; but I was continually struck with surprise how closely the three natives on board H. M. S. *Beagle,* who had lived some years in England, and could talk a little English, resembled us in disposition and in most of our mental faculties. If no organic being excepting man had possessed any mental power, or if his powers had been of a wholly different nature from those of the lower animals, then we should never have been able to convince ourselves that our high faculties had been gradually developed. But it can be shown that there is no fundamental difference of this kind. We must also admit that there is a much wider interval in mental power between one of the lowest fishes, as a lamprey or lancelet, and one of the higher apes, than between an ape and man; yet this interval is filled up by numberless gradations.

Most of the more complex emotions are common to the higher animals and ourselves. Everyone has seen how jealous a dog is of his master's affection, if lavished on any other creature; and I have observed the same fact with monkeys. This shows that animals not only love, but have desire to be loved. Animals manifestly feel emulation. They love approbation or praise; and a dog carrying a basket for his master exhibits in a high degree self-complacency or pride. There can, I think, be no doubt that a dog feels shame, as distinct from fear, and something very like modesty when begging too often for food. A great dog scorns the snarling of a little dog, and this may be called magnanimity. Several observers have stated that monkeys certainly dislike being laughed at; and they sometimes invent imaginary offenses. In the Zoological Gardens I saw a baboon who always got into a furious rage when his keeper took out a letter or book and read it aloud to him; and his rage was so violent that, as I witnessed on one occasion, he bit his own leg till the blood flowed. Dogs show what may be fairly called a sense of humor, as distinct from mere play; if a bit of stick or other such object be thrown to one, he will often carry it away for a short distance; and then squatting down with it on the ground close before him, will wait until his master comes quite close to take it away. The dog will then seize it and rush away in triumph, repeating the same maneuver, and evidently enjoying the practical joke.

We will now turn to the more intellectual emotions and faculties, which are very important, as forming the basis for the development of the higher mental powers. Animals manifestly enjoy excitement, and suffer from ennui, as may be seen with dogs, and, according to Rengger, with monkeys. All animals feel *Wonder,* and many exhibit *Curiosity.* They sometimes suffer from this latter quality, as when the hunter plays antics and thus attracts them; I have witnessed this with deer, and so it is with the wary chamois, and with some kinds of wild ducks.

Our domestic dogs are descended from wolves and jackals, and though they may not have gained in cunning, and may have lost in wariness and suspicion, yet they have progressed in certain moral qualities, such as in affection, trustworthiness, temper, and probably in general intelligence.

It may be freely admitted that no animal is self-conscious, if by this term it is implied, that he reflects on such points, as whence he comes or whither he will go, or what is life and death, and so forth. But how can we feel sure that an old dog with an excellent memory and some power of imagination, as shown by his dreams, never reflects on his past pleasures or pains in the chase? And this would be a form of self-consciousness.

The habitual use of articulate language is, however, peculiar to man; but he uses, in common with the lower animals, inarticulate cries to express his meaning, aided by gestures and the movements of the muscles of the face. This especially holds good with the more simple and vivid feelings, which are but little connected with our higher intelligence. Our cries of pain, fear, surprise, anger, together with their appropriate actions, and the murmur of a mother to her beloved child, are more expressive than any words. That which distinguishes man from the lower animals is not the understanding of articulate sounds, for, as everyone knows, dogs understand many words and sentences. In this respect they are at the same stage of development as infants, between the ages of ten and twelve months, who understand many words and short sentences, but cannot yet utter a single word. It is not the mere articulation which is our distinguishing character, for parrots and other birds possess this power. . . . The lower animals differ from man solely in his almost infinitely larger power of associating together the most diversified sounds and ideas; and this obviously depends on the high development of his mental powers.

[The sense of beauty] has been declared to be peculiar to man. I refer here only to the pleasure given by certain colors, forms, and sounds, and which may fairly be called a sense of the beautiful; with cultivated men such sensations are, however, intimately associated with complex ideas and trains of thought. When we behold a male bird elaborately displaying his graceful plumes or splendid colors before the female, whilst other birds, not thus decorated, make no such display, it is impossible to doubt that she admires the beauty of her male partner. As women everywhere deck themselves with these plumes, the beauty of such ornaments cannot be disputed. As we shall see later, the nests of humming-birds, and the playing passages of bower-birds are tastefully ornamented with gaily-colored objects; and this shows that they must receive some kind of pleasure from the sight of such things. With the great majority of animals, however, the taste for the beautiful is confined, as far as we can judge, to the attractions of the opposite sex.

The feeling of religious devotion is a highly complex one, consisting of love, complete submission to an exalted and mysterious superior, a strong sense of dependence, fear, reverence, gratitude, hope for the future, and perhaps other elements. No being could experience so complex an emotion until advanced in his intellectual and moral faculties to at least a moderately high level. Nevertheless, we see some distant approach to this state of mind in the deep love of a dog for his master, associated with complete submission, some fear, and perhaps other feelings. The behavior of a dog when returning to his master after an absence, and, as I may add, of a monkey to his beloved keeper, is widely different from that towards their fellows.

Conscience and Morality

The following proposition seems to me in a high degree probable—namely, that any animal whatever, endowed with well-marked social instincts, the parental and filial affections being here included, would inevitably acquire a moral sense or conscience, as soon as its intellectual powers had become as well, or nearly as well developed, as in man. For, *firstly,* the social instincts lead an animal to take pleasure in the society of its fellows, to feel a certain amount of sympathy with them, and to perform various services for them. The services may be of a definite and evidently instinctive nature; or there may be only a wish and readiness, as with most of the higher social animals, to aid their fellows in certain general ways. . . . *Secondly,* as soon as the mental faculties had become highly developed, images of all past actions and motives would be incessantly passing through the brain of each individual: and that feeling of dissatisfaction, or even misery, which invariably results, as we shall hereafter see, from any unsatisfied instinct, would arise, as often as it was perceived that the enduring and always present social instinct had yielded to some other instinct, at the time stronger, but neither enduring in its nature, nor leaving behind it a very vivid impression. It is clear that many instinctive desires, such as that of hunger, are in their nature of short duration; and after being satisfied, are not readily or vividly recalled. *Thirdly,* after the power of language had been acquired, and the wishes of the community could be expressed, the common opinion how each member ought to act for the public good, would naturally become in a paramount degree the guide to action. But it should be borne in mind that however great weight we may attribute to public opinion, our regard for the approbation and disapprobation of our fellows depends on sympathy, which, as we shall see, forms an essential part of the social instinct, and is indeed its foundation-stone. *Lastly,* habit in the individual would ultimately play a very important part in guiding the conduct of each member; for the social instinct, together with sympathy, is, like any other instinct, greatly strengthened by habit, and so consequently would be obedience to the wishes and judgment of the community.

We have not, however, as yet considered the main point, on which, from our present point of view, the whole question of the moral sense turns. Why should a man feel that he ought to obey one instinctive desire rather than another? Why is he bitterly regretful, if he has yielded to a strong sense of self-preservation, and has not risked his life to save that of a fellow-creature? Or why does he regret having stolen food from hunger?

It is evident in the first place, that with mankind the instinctive impulses have different degrees of strength; a savage will risk his own life to save that of a member of the same community, but will be wholly indifferent about a stranger: a young and timid mother urged by the maternal instinct will, without a moment's hesitation, run the greatest danger for her own infant, but not for a mere fellow-creature. Nevertheless many a civilized man, or even boy, who never before risked his life for another, but full of courage and sympathy, has disregarded the instinct of self-preservation, and plunged at once into a torrent to save a drowning man, though a stranger. In this case man is impelled by the same instinctive motive, which made the heroic little American monkey, formerly described, save his keeper, by attacking the great and dreaded baboon. Such actions as the above appear to be the simple result of the greater strength of the social or maternal instincts than that of any other instinct or motive; for they are performed too instantaneously for reflection, or for pleasure or pain to be felt at the time; though, if prevented by any cause, distress or even misery might be felt.

A man cannot prevent past impressions often repassing through his mind; he will thus be driven to make a comparison between the impressions of past hunger, vengeance satisfied, or danger shunned at other men's cost, with the almost ever-present instinct of sympathy, and with his early knowledge of what others consider as praiseworthy or blameable. This knowledge cannot be banished from his mind, and from instinctive sympathy is esteemed of great moment. He will then feel as if he had been balked in following a present instinct or habit, and this with all animals causes dissatisfaction, or even misery.

As man advances in civilization, and small tribes are united into larger communities, the simplest reason would tell each individual that he ought to extend his social instincts and sympathies to all the members of the same nation, though personally unknown to him. This point being once reached, there is only an artificial barrier to prevent his sympathies extending to the men of all nations and races. If, indeed, such men are separated from him by great differences in appearance or habits, experience unfortunately shows us how long it is, before we look at them as our fellow-creatures. Sympathy beyond the confines of man, that is, humanity to the lower animals, seems to be one of the latest moral acquisitions. It is apparently unfelt by savages, except towards their pets. How little the old Romans knew of it is shown by their abhorrent gladiatorial exhibitions. The very idea of humanity, as far as I could observe, was new to most of the Gauchos of the Pampas. This virtue, one of the noblest with which man is endowed, seems to arise incidentally from our sympathies becoming more tender and more widely diffused, until they are extended to all sentient beings. As soon as this virtue is honored and practiced by some few men, it spreads through instruction and example to the young, and eventually becomes incorporated in public opinion.

There is not the least inherent improbability, as it seems to me, in virtuous tendencies being more or less strongly inherited; for, not to mention the various dispositions and habits transmitted by many of our domestic animals to their offspring, I have heard of authentic cases in which a desire to steal and a tendency to lie appeared to run in families of the upper ranks; and as stealing is a rare crime in the wealthy classes, we can hardly account by accidental coincidence for the tendency occurring in two or three members of the same family. If bad tendencies are transmitted, it is probable that good ones are likewise transmitted. That the state of the body by affecting the brain, has great influence on the moral tendencies is known to most of those who have suffered from chronic derangements of the digestion or liver.

Finally the social instincts, which no doubt were acquired by man as by the lower animals for the good of the community, will

The Natural Evolution of Man and Morality

from the first have given to him some wish to aid his fellows, some feeling of sympathy, and have compelled him to regard their approbation and disapprobation. Such impulses will have served him at a very early period as a rude rule of right and wrong. But as man gradually advanced in intellectual power, and was enabled to trace the more remote consequences of his actions; as he acquired sufficient knowledge to reject baneful customs and superstitions; as he regarded more and more, not only the welfare, but the happiness of his fellow-men; as from habit, following on beneficial experience, instruction and example, his sympathies became more tender and widely diffused, extending to men of all races, to the imbecile, maimed, and other useless members of society, and finally to the lower animals,—so would the standard of his morality rise higher and higher.

After having yielded to some temptation we feel a sense of dissatisfaction, shame, repentance, or remorse, analogous to the feelings caused by other powerful instincts or desires, when left unsatisfied or balked. We compare the weakened impression of a past temptation with the ever present social instincts, or with habits, gained in early youth and strengthened during our whole lives, until they have become almost as strong as instincts. If with the temptation still before us we do not yield, it is because either the social instinct or some custom is at the moment predominant, or because we have learnt that it will appear to us hereafter the stronger, when compared with the weakened impression of the temptation, and we realize that its violation would cause us suffering. Looking to future generations, there is no cause to fear that the social instincts will grow weaker, and we may expect that virtuous habits will grow stronger, becoming perhaps fixed by inheritance. In this case the struggle between our higher and lower impulses will be less severe, and virtue will be triumphant.

Summary. There can be no doubt that the difference between the mind of the lowest man and that of the highest animal is immense. An anthropomorphous ape, if he could take a dispassionate view of his own case, would admit that though he could form an artful plan to plunder a garden—though he could use stones for fighting or for breaking open nuts, yet that the thought of fashioning a stone into a tool was quite beyond his scope. Still less, as he would admit, could he follow out a train of metaphysical reasoning, or solve a mathematical problem, or reflect on God, or admire a grand natural scene. Some apes, however, would probably declare that they could and did admire the beauty of the colored skin and fur of their partners in marriage. They would admit, that though they could make other apes understand by cries some of their perceptions and simpler wants, the notion of expressing definite ideas by definite sounds had never crossed their minds. They might insist that they were ready to aid their fellow-apes of the same troop in many ways, to risk their lives for them, and to take charge of their orphans; but they would be forced to acknowledge that disinterested love for all living creatures, the most noble attribute of man, was quite beyond their comprehension.

Nevertheless the difference in mind between man and the higher animals, great as it is, certainly is one of degree and not of kind. We have seen that the senses and intuitions, the various emotions and faculties, such as love, memory, attention, curiosity, imitation, reason, etc., of which man boasts, may be found in an incipient, or even sometimes in a well-developed condition, in the lower animals. They are also capable of some inherited improvement, as we see in the domestic dog compared with the wolf or jackal. If it could be proved that certain high mental powers, such as the formation of general concepts, self-consciousness, etc., were absolutely peculiar to man, which seems extremely doubtful, it is not improbable that these qualities are merely the incidental results of other highly advanced intellectual faculties; and these again mainly the result of the continued use of a perfect language. At what age does the newborn infant possess the power of abstraction, or become self-conscious, and reflect on its own existence? We cannot answer; nor can we answer in regard to the ascending organic scale. The half-art, half-instinct of language still bears the stamp of its gradual evolution. The ennobling belief in God is not universal with man; and the belief in spiritual agencies naturally follows from other mental powers. The moral sense perhaps affords the best and highest distinction between man and the lower animals; but I need say nothing on this head, as I have so lately endeavored to show that the social instincts—the prime principle of man's moral constitution—with the aid of active intellectual powers and the effects of habit, naturally lead to the golden rule, "As ye would that men should do to you, do ye to them likewise"; and this lies at the foundation of morality.

In the next chapter I shall make some few remarks on the probable steps and means by which the several mental and moral faculties of man have been gradually evolved. That such evolution is at least possible, ought not to be denied, for we daily see these faculties developing in every infant; and we may trace a perfect gradation from the mind of an utter idiot, lower than that of an animal low in the scale, to the mind of a Newton.

Human Genealogy

Even if it be granted that the difference between man and his nearest allies is as great in corporeal structure as some naturalists maintain, and although we must grant that the difference between them is immense in mental power, yet the facts given in the earlier chapters appear to declare, in the plainest manner, that man is descended from some lower form, notwithstanding that connecting-links have not hitherto been discovered.

The greater number of naturalists who have taken into consideration the whole structure of man, including his mental faculties, have followed Blumenbach and Cuvier, and have placed man in a separate Order, under the title of the Bimana, and therefore on an equality with the orders of the Quadrumana, Carnivora, etc. Recently many of our best naturalists have recurred to the view first propounded by Linnaeus, so remarkable for his sagacity, and have placed man in the same Order with the Quadrumana, under the title of the Primates. The justice of this conclusion will be admitted: for in the first place, we must bear in mind the comparative insignificance for classification of the great development of the brain in man, and that the strongly marked differences between the skulls of man and the Quadrumana . . . apparently follow from their differently developed brains. In the second place, we must remember that nearly all the other and more important differences between man and the Quadrumana are manifestly adaptive in their nature, and relate chiefly to the erect position of man; such as the structure of his hand, foot, and pelvis, the curvature of his spine, and the position of his head.

In the class of mammals the steps are not difficult to conceive which led from the ancient Monotremata to the ancient Marsupials; and from these to the early progenitors of the placental mammals. We may thus ascend to the Lemuridae; and the interval is not very wide from these to the Simiadae. The Simiadae then branched off into two great stems, the New World and Old World monkeys; and from the latter, at a remote period, Man, the wonder and glory of the Universe, proceeded.

Thus we have given to man a pedigree of prodigious length, but not, it may be said, of noble quality. The world, it has often been remarked, appears as if it had long been preparing for the advent of man: and this, in one sense is strictly true, for he owes his birth to a long line of progenitors. If any single link in this chain had never existed, man would not have been exactly what he now is. Unless we wilfully close our eyes, we may, with our present knowledge, approximately recognize our parentage; nor need we feel ashamed of it. The most humble organism is something much higher than the inorganic dust under our feet; and no one with an unbiased mind can study any living creature, however humble, without being struck with enthusiasm at its marvellous structure and properties.

Alfred Russel Wallace
THE INADEQUACIES OF DARWINIAN EVOLUTION

Like Darwin, Alfred Russel Wallace (1823–1913) was already convinced that species had evolved from prior ones before he was struck one day with the idea that the method behind evolution was selection by natural competition. Within two days Wallace had written an abstract of his idea for publication and mailed it off to Charles Darwin. The two naturalists soon introduced their theory to the world through reading joint papers before the Linnaean Society in London in July 1858. As is clear in the reading below, Wallace remained an enthusiastic evolutionist all his life, and believed that it was "inconceivable" that man also had not received his bodily structure from prior mammalian ancestors. But similar to the great, influential geologist Charles Lyell (1797–1875), Wallace believed that something that deified or transcended scientific explanation had to be at work in the organic world. How could natural selection, asked Wallace and Lyell, produce a being (man) who is no longer necessarily bound and subject to environmental changes? How could nature produce an organic being who is superior to itself in ingenuity, aesthetic sensibility and creative, mental power? As a spiritualist interested in psychic phenomena, Wallace sought for explanations to this vitalism in the "unseen world" of spirit.

To any one who considers the structure of man's body, even in the most superficial manner, it must be evident that it is the body of an animal, differing greatly, it is true, from the bodies of all other animals, but agreeing with them in all essential features. The bony structure of man classes him as a vertebrate; the mode of suckling his young classes him as a mammal; his blood, his muscles, and his nerves, the structure of his heart with its veins and arteries, his lungs and his whole respiratory and circulatory systems, all closely correspond to those of other mammals, and are often almost identical with them. He possesses the same number of limbs terminating in the same number of digits as belong fundamentally to the mammalian class. His senses are identical with theirs, and his organs of sense are the same in number and occupy the same relative position. Every detail of structure which is common to the mammalia as a class is found also in man, while he only differs from them in such

Abridged from Alfred Russel Wallace, *Darwinism* (London: Macmillan, 1891).

ways and degrees as the various species or groups of mammals differ from each other. If, then, we have good reason to believe that every existing group of mammalia has descended from some common ancestral form—as we saw to be so completely demonstrated in the case of the horse tribe—and that each family, each order, and even the whole class must similarly have descended from some much more ancient and more generalized type, it would be in the highest degree improbable—so improbable as to be almost inconceivable—that man, agreeing with them so closely in every detail of his structure, should have had some quite distinct mode of origin.

The Origin of the Moral and Intellectual Nature of Man

. . . I fully accept Mr. Darwin's conclusion as to the essential identity of man's bodily structure with that of the higher mammalia, and his descent from some ancestral form common to man and the anthropoid apes. The evidence of such descent appears to me to be overwhelming and conclusive. Again, as to the cause and method of such descent and modification, we may admit, at all events provisionally, that the laws of variation and natural selection, acting through the struggle for existence and the continual need of more perfect adaptation to the physical and biological environments, may have brought about, first that perfection of bodily structure in which he is so far above all other animals, and in coordination with it the larger and more developed brain, by means of which he has been able to utilize that structure in the more and more complete subjection of the whole animal and vegetable kingdoms to his service.

But this is only the beginning of Mr. Darwin's work, since he goes on to discuss the moral nature and mental faculties of man, and derives these too by gradual modification and development from the lower animals. Although, perhaps, nowhere distinctly formulated, his whole argument tends to the conclusion that man's entire nature and all his faculties, whether moral, intellectual, or spiritual, have been derived from their rudiments in the lower animals, in the same manner and by the action of the same general laws as his physical structure has been derived. As this conclusion appears to me not to be supported by adequate evidence, and to be directly opposed to many well-ascertained facts, I propose to devote a brief space to its discussion.

The Inadequacies of Darwinian Evolution

The Argument from Continuity

Mr. Darwin's mode of argument consists in showing that the rudiments of most, if not of all, the mental and moral faculties of man can be detected in some animals. The manifestations of intelligence, amounting in some cases to distinct acts of reasoning, in many animals, are adduced as exhibiting in a much less degree the intelligence and reason of man. Instances of curiosity, imitation, attention, wonder, and memory are given; while examples are also adduced which may be interpreted as proving that animals exhibit kindness to their fellows, or manifest pride, contempt, and shame. Some are said to have the rudiments of language, because they utter several different sounds, each of which has a definite meaning to their fellows or to their young; others the rudiments of arithmetic, because they seem to count and remember up to three, four, or even five. A sense of beauty is imputed to them on account of their own bright colors or the use of colored objects in their nests; while dogs, cats, and horses are said to have imagination, because they appear to be disturbed by dreams. Even some distant approach to the rudiments of religion is said to be found in the deep love and complete submission of a dog to his master.

Turning from animals to man, it is shown that in the lowest savages many of these faculties are very little advanced from the condition in which they appear in the higher animals; while others, although fairly well exhibited, are yet greatly inferior to the point of development they have reached in civilized races. In particular, the moral sense is said to have been developed from the social instincts of savages, and to depend mainly on the enduring discomfort produced by any action which excites the general disapproval of the tribe.

The question of the origin and nature of the moral sense and of conscience is far too vast and complex to be discussed here, and a reference to it has been introduced only to complete the sketch of Mr. Darwin's view of the continuity and gradual development of all human faculties from the lower animals up to savages, and from savage up to civilized man. The point to which I wish specially to call attention is, that to prove continuity and the progressive development of the intellectual and moral faculties from animals to man, is not the same as proving that these faculties have been developed by natural selection; and this last is what Mr. Darwin has hardly attempted, although to support his theory it was absolutely essential to prove it. Because man's physical structure has been developed from an animal form by natural selection, it does not necessarily follow that his mental nature, even though developed *pari passu* with it, has been developed by the same causes only.

It is not, therefore, to be assumed, without proof or against independent evidence, that the later stages of an apparently continuous development are necessarily due to the same causes only as the earlier stages. Applying this argument to the case of man's intellectual and moral nature, I propose to show that certain definite portions of it could not have been developed by variation and natural selection alone, and that, therefore, some other influence, law, or agency is required to account for them. If this can be clearly shown for any one or more of the special faculties of intellectual man, we shall be justified in assuming that the same unknown cause or power may have had a much wider influence, and may have profoundly influenced the whole course of his development.

The Origin of the Mathematical Faculty

We have ample evidence that, in all the lower races of man, what may be termed the mathematical faculty is, either absent, or, if present, quite unexercised. The Bushmen and the Brazilian Wood-Indians are said not to count beyond two. Many Australian tribes only have words for one and two, which are combined to make three, four, five, or six, beyond which they do not count.

When we turn to the more civilized races, we find the use of numbers and the art of counting greatly extended. Even the Tongas of the South Sea islands are said to have been able to count as high as 100,000. But mere counting does not imply either the possession or the use of anything that can be really called the mathematical faculty, the exercise of which in any broad sense has only been possible since the introduction of the decimal notation. . . . It is, however, during the last three centuries only that the civilized world appears to have become conscious of the possession of a marvelous faculty which, when supplied with the necessary tools in the decimal notation, the elements of algebra and geometry, and the power of rapidly communicating discoveries and ideas by the art of printing,

The Inadequacies of Darwinian Evolution

has developed to an extent, the full grandeur of which can be appreciated only by those who have devoted some time (even if unsuccessfully) to the study.

The facts now set forth as to the almost total absence of mathematical faculty in savages and its wonderful development in quite recent times, are exceedingly suggestive, and in regard to them we are limited to two possible theories. Either prehistoric and savage man did not possess this faculty at all (or only in its merest rudiments); or they did possess it, but had neither the means nor the incitements for its exercise.

We have to ask, therefore, what relation the successive stages of improvement of the mathematical faculty had to the life or death of its possessors; to the struggles of tribe with tribe, or nation with nation; or to the ultimate survival of one race and the extinction of another. If it cannot possibly have had any such effects, then it cannot have been produced by natural selection.

It is evident that in the struggles of savage man with the elements and with wild beasts, or of tribe with tribe, this faculty can have had no influence. It had nothing to do with the early migrations of man, or with the conquest and extermination of weaker by more powerful peoples. The Greeks did not successfully resist the Persian invaders by any aid from their few mathematicians, but by military training, patriotism, and self-sacrifice. The barbarous conquerors of the East, Timurlane and Genghis Khan, did not owe their success to any superiority of intellect or of mathematical faculty in themselves or their followers. Even if the great conquests of the Romans were, in part, due to their systematic military organization, and to their skill in making roads and encampments, which may, perhaps, be imputed to some exercise of the mathematical faculty, that did not prevent them from being conquered in turn by barbarians, in whom it was almost entirely absent. . . . We conclude, then, that the present gigantic development of the mathematical faculty is wholly unexplained by the theory of natural selection, and must be due to some altogether distinct cause.

The Origin of the Musical and Artistic Faculties

As with the mathematical, so with the musical faculty, it is impossible to trace any connection between its possession and survival in the struggle for existence. It seems to have arisen as a *result* of social and intellectual advancement, not as a *cause;* and there is some evidence that it is latent in the lower races, since under European training native military bands have been formed in many parts of the world, which have been able to perform creditably the best modern music.

The artistic faculty has run a somewhat different course, though analogous to that of the faculties already discussed. Most savages exhibit some rudiments of it, either in drawing or carving human or animal figures; but, almost without exception, these figures are rude and such as would be executed by the ordinary inartistic child. In fact, modern savages are, in this respect hardly equal to those prehistoric men who represented the mammoth and the reindeer on pieces of horn or bone. With any advance in the arts of social life, we have a corresponding advance in artistic skill and taste, rising very high in the art of Japan and India, but culminating in the marvelous sculpture of the best period of Grecian history.

These several developments of the artistic faculty, whether manifested in sculpture, painting, or architecture, are evidently outgrowths of the human intellect which have no immediate influence on the survival of individuals or of tribes, or on the success of nations in their struggles for supremacy or for existence. The glorious art of Greece did not prevent the nation from falling under the sway of the less advanced Roman; while we ourselves, among whom art was the latest to arise, have taken the lead in the colonization of the world, thus proving our mixed race to be the fittest to survive.

We have thus shown, by two distinct lines of argument, that faculties are developed in civilized man which, both in their mode of origin, their function, and their variations, are altogether distinct from those other characters and faculties which are essential to him, and which have been brought to their actual state of efficiency by the necessities of his existence. And besides the three which have been specially referred to, there are others which evidently belong to the same class. Such is the metaphysical faculty, which enables us to form abstract conceptions of a kind the most remote from all practical applications, to discuss the ultimate causes of things, the nature and qualities of matter, motion, and force, of space and time, of cause and effect, of will and conscience.

The Interpretation of the Facts

The facts now set forth prove the existence of a number of mental faculties which either do not exist at all or exist in a very rudimentary condition in savages, but appear almost suddenly and in perfect development in the higher civilized races. These same faculties are further characterized by their sporadic character, being well developed only in a very small proportion of the community; and by the enormous amount of variation in their development, the higher manifestations of them being many times—perhaps a hundred or a thousand times—stronger than the lower. Each of these characteristics is totally inconsistent with any action of the law of natural selection in the production of the faculties referred to; and the facts, taken in their entirety, compel us to recognize some origin for them wholly distinct from that which has served to account for the animal characteristics—whether bodily or mental—of man.

The special faculties we have been discussing clearly point to the existence in man of something which he has not derived from his animal progenitors—something which we may best refer to as being of a spiritual essence or nature, capable of progressive development under favorable conditions. On the hypothesis of this spiritual nature, superadded to the animal nature of man, we are able to understand much that is otherwise mysterious or unintelligible in regard to him, especially the enormous influence of ideas, principles, and beliefs over his whole life and actions. Thus alone we can understand the constancy of the martyr, the unselfishness of the philanthropist, the devotion of the patriot, the enthusiasm of the artist, and the resolute and persevering search of the scientific worker after nature's secrets. Thus we may perceive that the love of truth, the delight in beauty, the passion for justice, and the thrill of exultation with which we hear of any act of courageous self-sacrifice, are the workings within us of a higher nature which has not been developed by means of the struggle for material existence.

It will, no doubt, be urged that the admitted continuity of man's progress from the brute does not admit of the introduction of new causes, and that we have no evidence of the sudden change of nature which such introduction would bring about. The fallacy as to new causes involving any breach of continuity, or any sudden or abrupt change, in the effects, has already been shown; but we will further point out that there are at least three stages in the development of the organic world when some new cause or power must necessarily have come into action.

The first stage is the change from inorganic to organic, when the earliest vegetable cell, or the living protoplasm out of which it arose, first appeared. This is often imputed to a mere increase of complexity of chemical compounds; but increase of complexity, with consequent instability, even if we admit that it may have produced protoplasm as a chemical compound, could certainly not have produced *living* protoplasm—protoplasm which has the power of growth and of reproduction, and of that continuous process of development which has resulted in the marvelous variety and complex organization of the whole vegetable kingdom. There is in all this something quite beyond and apart from chemical changes, however complex; and it has been well said that the first vegetable cell was a new thing in the world, possessing altogether new powers—that of extracting and fixing carbon from the carbon-dioxide of the atmosphere, that of indefinite reproduction, and, still more marvelous, the power of variation and of reproducing those variations till endless complications of structure and varieties of form have been the result. Here, then, we have indications of a new power at work, which we may term *vitality,* since it gives to certain forms of matter all those characters and properties which constitute Life.

The next stage is still more marvelous, still more completely beyond all possibility of explanation by matter, its laws and forces. It is the introduction of sensation or consciousness, constituting the fundamental distinction between the animal and vegetable kingdoms. Here all idea of mere complication of structure producing the result is out of the question. We feel it to be altogether preposterous to assume that at a certain stage of complexity of atomic constitution, and as a necessary result of that complexity alone, an *ego* should start into existence, a thing that *feels,* that is *conscious* of its own existence. Here we have the certainty that something new has arisen, a being whose nascent consciousness has gone on increasing in power and definiteness till it has culminated in the higher animals. No verbal explanation or attempt at explanation—such as the statement that life is the result of the molecular forces of the protoplasm, or that the whole existing organic universe from

The Inadequacies of Darwinian Evolution

the amoeba up to man was latent in the fire-mist from which the solar system was developed—can afford any mental satisfaction, or help us in any way to a solution of the mystery.

The third stage is, as we have seen, the existence in man of a number of his most characteristic and noblest faculties, those which raise him furthest above the brutes and open up possibilities of almost indefinite advancement. These faculties could not possibly have been developed by means of the same laws which have determined the progressive development of the organic world in general, and also of man's physical organism.

These three distinct stages of progress from the inorganic world of matter and motion up to man, point clearly to an unseen universe—to a world of spirit, to which the world of matter is altogether subordinate. To this spiritual world we may refer the marvelously complex forces which we know as gravitation, cohesion, chemical force, radiant force, and electricity, without which the material universe could not exist for a moment in its present form, and perhaps not at all, since without these forces, and perhaps others which may be termed atomic, it is doubtful whether matter itself could have any existence. And still more surely can we refer to it those progressive manifestations of Life in the vegetable, the animal, and man—which we may classify as unconscious, conscious, and intellectual life—and which probably depend upon different degrees of spiritual influx.

Concluding Remarks

Those who admit my interpretation of the evidence now adduced—strictly scientific evidence in its appeal to facts which are clearly what ought *not* to be on the materialistic theory—will be able to accept the spiritual nature of man, as not in any way inconsistent with the theory of evolution, but as dependent on those fundamental laws and causes which furnish the very materials for evolution to work with. They will also be relieved from the crushing mental burden imposed upon those who—maintaining that we, in common with the rest of nature, are but products of the blind eternal forces of the universe, and believing also that the time must come when the sun will lose his heat and all life on the earth necessarily cease—have to contemplate a not very distant future in which all this glorious earth—which for untold millions of years has been slowly developing forms of life and beauty to culminate at last in man—shall be as if it had never existed; who are compelled to suppose that all the slow growths of our race struggling towards a higher life, all the agony of martyrs, all the groans of victims, all the evil and misery and undeserved suffering of the ages, all the struggles for freedom, all the efforts towards justice, all the aspirations for virtue and the well-being of humanity, shall absolutely vanish, and, "like the baseless fabric of a vision, leave not a wrack behind."

As contrasted with this hopeless and soul-deadening belief, we, who accept the existence of a spiritual world, can look upon the universe as a grand consistent whole adapted in all its parts to the development of spiritual beings capable of indefinite life and perfectibility. To us, the whole purpose, the only *raison d'être* of the world—with all its complexities of physical structure, with its grand geological progress, the slow evolution of the vegetable and animal kingdoms, and the ultimate appearance of man—was the development of the human spirit in association with the human body. From the fact that the spirit of man—the man himself—*is* so developed, we may well believe that this is the only, or at least the best, way for its development; and we may even see in what is usually termed "evil" on the earth, one of the most efficient means of its growth. For we know that the noblest faculties of man are strengthened and perfected by struggle and effort; it is by unceasing warfare against physical evils and in the midst of difficulty and danger that energy, courage, self-reliance, and industry have become the common qualities of the northern races; it is by the battle with moral evil in all its hydra-headed forms, that the still nobler qualities of justice and mercy and humanity and self-sacrifice have been steadily increasing in the world. Beings thus trained and strengthened by their surroundings, and possessing latent faculties capable of such noble development, are surely destined for a higher and more permanent existence; and we may confidently believe with our greatest living poet—

> *That life is not as idle ore,*
> *But iron dug from central gloom,*
> *And heated hot with burning fears,*
> *And dipt in baths of hissing tears,*

And batter'd with the shocks of doom
To shape and use.[1]

We thus find that the Darwinian theory, even when carried out to its extreme logical conclusion, not only does not oppose, but lends a decided support to, a belief in the spiritual nature of man. It shows us how man's body may have been developed from that of a lower animal form under the law of natural selection; but it also teaches us that we possess intellectual and moral faculties which could not have been so developed, but must have had another origin; and for this origin we can only find an adequate cause in the unseen universe of Spirit.

---

[1] Alfred Lord Tennyson, "In Memoriam A. H. H."—Ed.

John Fiske
THE ASCENT OF MAN

As a precocious undergraduate at Harvard, John Fiske (1823–1913) jeopardized his chances for graduation by his enthusiastic advocacy of Darwinian evolution in the then conservative Harvard environment. Nevertheless, he graduated in 1863 and began successfully and brilliantly to popularize Darwinism in America. Like Alfred Russel Wallace, Fiske believed that Darwin and Huxley had made evolution too limiting and mechanical. Focusing also on man and mind, Fiske interpreted evolution as a cosmic process by which inorganic matter brought into being organic forms, and organic forms through the development of mental faculties and social systems became progressively noble and moral. Evolution, Fiske believed, should thus be regarded as the ascent of organic nature through man—and beyond—rather than a process that confines an analysis of mankind to physical processes stripped of transcendent meaning. In this sense Fiske serves as a "precursor" to the mystical evolutionary ideas of Teilhard de Chardin.

First Stages in the Genesis of Man

Let us begin by drawing a correct though slight outline sketch of what the cosmic process of evolution has been. It is not strange that when biologists speak of evolution they should often or usually have in mind simply the modifications wrought in plants and animals by means of natural selection. For it was by calling attention to such modifications that Darwin discovered a true cause of the origin of species by physiological descent from allied species. Thus was demonstrated the fact of evolution in its most important province; men of science were convinced that the higher forms of life are derived from lower forms, and the old notion of special creations was exploded once and forever. This was a great scientific achievement, one of the greatest known to history, and it is therefore not strange that language should often be employed as if Evolutionism and Darwinism were synonymous. Yet not only are there extensive regions in the doctrine of evolution about which Darwin knew very little, but even as regards the genesis of species his theory was never developed in his own hands so far as to account satisfactorily for the genesis of man.

Abridged from John Fiske, *Through Nature to God* (Boston: Houghton Mifflin, 1899).

It must be borne in mind that while the natural selection of physical variations will go far toward explaining the characteristics of all the plants and all the beasts in the world, it remains powerless to account for the existence of man. Natural selection of physical variations might go on for a dozen eternities without any other visible result than new forms of plant and beast in endless and meaningless succession. The physical variations by which man is distinguished from apes are not great. His physical relationship with the ape is closer than that between cat and dog, which belong to different families of the same order; it is more like that between cat and leopard, or between dog and fox, different genera in the same family. But the moment we consider the minds of man and ape, the gap between the two is immeasurable. Mr. Mivart has truly said that, with regard to their total value in nature, the difference between man and ape transcends the difference between ape and blade of grass. I should be disposed to go further and say, that while for zoological man you can hardly erect a distinct family from that of the chimpanzee and orang, on the other hand, for psychological man you must erect a distinct kingdom; nay, you must even dichotomize the universe, putting man on one side and all things else on the other. How can this overwhelming contrast between psychical and physical difference be accounted for? The clue was furnished by Alfred Russel Wallace, the illustrious co-discoverer of natural selection. Wallace saw that along with the general development of mammalian intelligence a point must have been reached in the history of one of the primates, when variations of intelligence were more profitable to him than variations in body. From that time forth that primate's intelligence went on by slow increments acquiring new capacity, while his body changed but little. When once he could strike fire, and chip a flint, and use a club, and strip off the bear's hide to cover himself, there was clearly no further use in thickening his own hide, or lengthening and sharpening his claws. Natural selection is the keenest capitalist in the universe; she never loses an instant in seizing the most profitable place for investment, and her judgment is never at fault. Forthwith, for a million years or more she invested all her capital in the psychical variations of this favored primate, making little change in his body except so far as to aid in the general result, until by and by something like human intelligence of a low grade, like that of the Australian or the Andaman islander, was achieved. The genesis of humanity was by no means yet completed, but an enormous gulf had been crossed.

After throwing out this luminous suggestion Mr. Wallace never followed it up as it admitted and deserved. It is too much to expect one man to do everything, and his splendid studies in the geographical distribution of organisms may well have left him little time for work in this direction. Who can fail to see that the selection of psychical variations, to the comparative neglect of physical variations, was the opening of a new and greater act in the drama of creation? Since that new departure the Creator's highest work has consisted not in bringing forth new types of body, but in expanding and perfecting the psychical attributes of the one creature in whose life those attributes have begun to acquire predominance. Along this human line of ascent there is no occasion for any further genesis of species, all future progress must continue to be not zoological, but psychological, organic evolution gives place to civilization. Thus in the long series of organic beings man is the last; the cosmic process, having once evolved this masterpiece, could thenceforth do nothing better than to perfect him.

The Central Fact in the Genesis of Man

This conclusion, which follows irresistibly from Wallace's theorem, that in the genesis of humanity natural selection began to follow a new path, already throws a light of promise over our whole subject, like the rosy dawn of a June morning. But the explanation of the genesis of humanity is still far from complete. If we compare man with any of the higher mammals, such as dogs and horses and apes, we are struck with several points of difference: *first,* the greatest progressiveness of man, the widening of the interval by which one generation may vary from its predecessor; *secondly,* the definite grouping in societies based on more or less permanent family relationships, instead of the indefinite grouping in miscellaneous herds or packs; *thirdly,* the possession of articulate speech; *fourthly,* the enormous increase in the duration of infancy, or the period when parental care is needed. . . . It is the prolonged infancy that has caused the progressiveness and the grouping into definite societies, while the development of language was a consequence of the increasing intelligence and sociality thus caused. In the genesis of humanity the central fact has been the increased duration of infancy. Now, can we assign for that increased duration an adequate cause? I think we can. The increase of intelligence is itself such a cause. A glance at the animal kingdom shows us no such thing as infancy among the lower orders. It is with warm-blooded birds and mammals that the phenomena of infancy and the correlative parental care really begin.

The Chief Cause of Man's Lengthened Infancy

The reason for this is that any creature's ability to perceive and to act depends upon the registration of experiences in his nerve-centers.

In writing, in walking, in talking, we are making use of nervous registrations that have been brought about by an accumulation of experiences. To pick up a pencil from the table may seem a very simple act, yet a baby cannot do it. It has been made possible only by the education of the eyes, of the muscles that move the eyes, of the arm and hand, and of the nerve-centers that coordinate one group of movements with another. All this multiform education has consisted in a gradual registration of experiences. In like manner all the actions of man upon the world about him are made up of movements, and every such movement becomes possible only when a registration is effected in sundry nerve-centers.

Presently the movements of limbs and sense organs come to be added, and as we rise in the animal scale, these movements come to be endlessly various and complex, and by and by implicate the nervous system more and more deeply in complex acts of perception, memory, reasoning, and volition. Obviously, therefore, in the development of the individual organism the demands of the nervous system upon the vital energies concerned in growth must come to be of paramount importance, and in providing for them the entire embryonic life must be most profoundly and variously affected. Though we may be unable to follow the processes in detail, the truth of this general statement is plain and undeniable.

I say, then, that when a creature's intelligence is low, and its experience very meager, consisting of a few simple perceptions and acts that occur throughout life with monotonous regularity, all the registration of this experience gets effected in the nerve-centers of its offspring before birth, and they come into the world fully equipped for the battle of life, like the snapping turtle, which snaps with decisive vigor as soon as it emerges from the egg. Nothing is left plastic to be finished after birth, and so the life of each generation is almost an exact repetition of its predecessor. But when a creature's intelligence is high, and its experience varied and complicated, the registration of all this experience in the nerve-centers of its offspring does not get accomplished before birth. . . .

We are now prepared to appreciate the marvelous beauty of nature's work in bringing man upon the scene. Nowhere is there any breach of continuity in the cosmic process. First we have natural selection at work throughout the organic world, bringing forth millions of species of plant and animal, seizing upon every advantage, physical or mental, that enables any species to survive in the universal struggle. So far as any outward observer, back in the Cretaceous or early Eocene periods, could surmise, this sort of confusion might go on forever. But all at once, perhaps somewhere in the upper Eocene or lower Miocene, it appears that among the primates, a newly developing family already distinguished for prehensile capabilities, one genus is beginning to sustain itself more by mental craft and shiftiness than by any physical characteristic. Forthwith does natural selection seize upon any and every advantageous variation in this craft and shiftiness, until this favored genus of primates, this *Homo Alalus,* or speechless man, as we may call him, becomes preeminent for sagacity.

The evidence is abundant that Homo Alalus, like his simian cousins, was a gregarious creature, and it is not difficult to see how, with increasing intelligence, the gestures and grunts used in the horde for signalling must come to be clothed with added associations of meaning, must gradually become generalized as signs of conceptions. This invention of spoken language, the first invention of nascent humanity, remains to this day its most fruitful invention. Henceforth ancestral experience could not simply be transmitted through its inheritable impress upon the nervous system, but its facts and lessons could become external materials and instruments of education. Then the children of Homo Alalus, no longer speechless, began to accumulate a fund of tradition, which in the fullness of time was to bloom forth in history and poetry, in science and theology.

I would emphasize the fact that nowhere do we find any breach of continuity, but one factor sets another in operation, which in turn reacts upon the first, and so on in a marvelously harmonious consensus. Surely if there is anywhere in the universe a story matchless for its romantic interest, it is the story of the genesis of man, now that we are at length beginning to be able to decipher it. We see that there is a good deal more in it than mere natural selection. At bottom, indeed, it is all a process of survival of the fittest, but the secondary agencies we have been considering have brought us to a point where our conception of the struggle for life must be enlarged. Out of the manifold compounding and recompounding of primordial clans have come the nations of mankind in various degrees of civilization, but already in the clan we find the ethical process at work. The clan has a code of morals well adapted to the conditions amid which it exists. There is an ethical sentiment in the clan; its members have duties toward it; it punishes sundry acts even with death, and rewards or extols sundry other acts. We are, in short, in an ethical atmosphere.

Love and Self-Sacrifice

Now the moment a man's voluntary actions are determined by conscious or unconscious reference to a standard outside of himself and his selfish motives, he has entered the world of ethics, he has begun to live in a moral atmosphere. Egoism has ceased to be all in all, and altruism . . . has begun to . . . assert its claim to sovereignty. . . . Nature's supreme end became the maintenance of the clan organization, the standard for the individual's conduct became shifted, permanently and forever shifted.

A word of caution may be needed. It is not for a moment to be supposed that when primitive men began crudely shaping their conduct with reference to a standard outside of self, they did so as the result of meditation, or with any realizing sense of what they were doing. That has never been the method of evolution. Its results steal upon the world noiselessly and unobserved, and only after they have long been with us does reason employ itself upon them.

The Cosmic Process Exists for Moral Ends

One of the most interesting, and in my opinion one of the most profoundly significant, facts in the whole process of evolution is the first appearance of religious sentiment at very nearly the same stage at which the moral law began to grow up. To the differential attributes of humanity already considered there needs to be added the possession of religious sentiment and religious ideas. We may safely say that this is the most important of all the distinctions between man and other animals; for to say so is simply to epitomize the whole of human experience as recorded in history, art, and literature. Along with the rise from gregariousness to incipient sociality, along with the first stammerings of articulate speech, along with the dawning discrimination between right and wrong, came the earliest feeble groping toward a world beyond that which greets the senses, the first dim recognition of the spiritual power that is revealed in and through the visible and palpable realm of nature. And universally since that time the notion of ethics has been inseparably associated with the notion of religion, and the sanction for ethics has been held to be closely related with the world beyond phenomena. There are philosophers who maintain that with the further progress of enlightenment this close relation will cease to be asserted, that ethics will be divorced from religion, and that the groping of the human soul after its god will be condemned as a mere survival from the errors of primitive savagery, a vain and idle reaching out toward a world of mere phantoms. I mention this opinion merely to express unqualified and total dissent from it. I believe it can be shown that one of the strongest implications of the doctrine of evolution is the everlasting reality of religion.

We have here been concerned purely with the ethical process itself, which we have found to be—as Huxley truly says in his footnote—part and parcel of the general process of evolution. Our historical survey of the genesis of humanity seems to show very forcibly that a society of human souls living in conformity to a perfect moral law is the end toward which, ever since the time when our solar system was a patch of nebulous vapor, the cosmic process has been aiming. After our cooling planet had become the seat of organic life, the process of natural selection went on for long ages seemingly, but not really at random; for our retrospect shows that its ultimate tendency was toward singling out one creature and exalting his intelligence.

The ethical process is not only part and parcel of the cosmic process, but it is its crown and consummation. Toward the spiritual perfection of humanity the stupendous momentum of the cosmic process has all along been tending. That spiritual perfection is the true goal of evolution, the divine end that was involved in the beginning. When Huxley asks us to believe that "the cosmic process has no sort of relation to moral ends," I feel like replying with the question, "Does not the cosmic process exist purely for the sake of moral ends?" Subtract from the universe its ethical meaning, and nothing remains but an unreal phantom, the figment of false metaphysics.

We have now arrived at a position from which a glimmer of light is thrown upon some of the dark problems connected with the moral government of the world. We can begin to see why misery and wrongdoing are permitted to exist, and why the creative energy advances by such slow and tortuous methods toward the fulfillment of its divine purpose. In order to understand these things, we must ask, What is the ultimate goal of the ethical process? According to the utilitarian philosophy that goal is the completion of human happiness. But this interpretation soon refutes itself. A world of completed happiness might well be a world of quiescence, of stagnation, of automatism, of blankness; the dynamics of evolution would have no place in it. But suppose we say that the ultimate goal of the ethical process is the perfecting of human character? This form of statement contains far more than the other. . . . The consummate product of a world of evolution is the character that *creates* happiness, that is replete with dynamic possibilities of fresh life and activity in directions forever new. Such a character is the reflected image of God, and in it are contained the promise and potency of life everlasting.

No such character could be produced by any act of special creation in a garden of eden. It must be the consummate efflorescence of long ages of evolution, and a world of evolution is necessarily characterized by slow processes, many of which to a looker-on seem like tentative experiments, with an enormous sacrifice of ephemeral forms of life. Thus while the Earth Spirit goes on, unhasting, yet unresting, weaving in the loom of time the visible garment of God, we begin to see that even what look like failures and blemishes have been from the outset involved in the accomplishment of the all-wise and all-holy purpose, the perfecting of the spiritual man in the likeness of his heavenly Father.

Reality of Religion

Now there was a critical moment in the history of our planet, when love was beginning to play a part hitherto unknown, when notions of right and wrong were germinating in the nascent human soul, when the family was coming into existence, when social ties were beginning to be knit, when winged words first took their flight through the air. It was the moment when the process of evolution was being shifted to a higher plane, when civilization was to be superadded to organic evolution, when the last and highest of creatures was coming upon the scene, when the dramatic purpose of creation was approaching fulfillment. At that critical moment we see the nascent human soul vaguely reaching forth toward something akin to itself not in the realm of fleeting phenomena but in the Eternal Presence beyond. An internal adjustment of ideas was achieved in correspondence with an unseen world. That the ideas were very crude and childlike, that they were put together with all manner of grotesqueness, is what might be expected. The cardinal fact is that the crude child-like mind was groping to put itself into relation with an ethical world not visible to the senses. And one aspect of this fact, not to be lightly passed over, is the fact that religion, thus ushered upon the scene coeval with the birth of humanity, has played such a dominant part in the subsequent evolution of human society that what history would be without it is quite beyond imagination.

The days of the antagonism between Science and religion must by and by come to an end. That antagonism has been chiefly due to the fact that religious ideas were until lately allied with the doctrine of special creations. They have therefore needed to be remodeled and considered from new points of view. But we have at length reached a stage where it is becoming daily more and more apparent that with the deeper study of nature the old strife between faith and knowledge is drawing to a close; and disentangled at last from that ancient slough of despond the human mind will breathe a freer air and enjoy a vastly extended horizon.

T. H. Huxley
SECULAR HUMANISM

The readings here display three aspects of T. H. Huxley's humanism. The excerpt from "Mr. Darwin's Critics" is taken from Huxley's response to the criticisms of Darwin's Descent of Man *by St. George Mivart and Alfred Russel Wallace. Opposing the position of Wallace given above, Huxley argued that the mental abilities of primitive humans did not greatly transcend their needs, and hence their mental aptitude could not be used as a proof that a spiritual, non-natural leap in evolution had occurred. Huxley, like Darwin, believed that mankind's evolution was purely natural and secular.*

The second excerpt, entitled "Man and the Lower Animals," is taken from Huxley's highly controversial Man's Place in Nature *(1863), in which he became the first Darwinian to affirm in print that man possessed "structural unity" with animals, "more particularly" with apes. In this reading Huxley directly faced the charge that his and Darwin's opinions "brutalized" mankind. Huxley maintained that man's functional and cultural superiority was a sufficient foundation for human dignity. The third excerpt, from* Science and Christian Tradition *(1895), shows that although Huxley regarded religion and morals as natural, evolutionary acquisitions, he believed that mankind's religious past—particularly as preserved in the Bible—contained cultural and moral insights that should be preserved, treasured and taught to the young.*

MR. DARWIN'S CRITICS

The gradual lapse of time has now separated us by more than a decade from the date of the publication of the *Origin of Species*—and whatever may be thought or said about Mr. Darwin's doctrines, or the manner in which he has propounded them, this much is certain, that, in a dozen years, the *Origin of Species* has worked as complete a revolution in biological science as the *Principia* did in astronomy—and it has done so, because, in the words of Helmholtz, it contains "an essentially new creative thought."

And as time has slipped by, a happy change has come over Mr. Darwin's critics. The mixture of ignorance and insolence which, at first, characterized a large proportion of the attacks with which he was assailed, is no longer the sad distinction of anti-Darwinian

Abridged from "Mr. Darwin's Critics" [1871] in T. H. Huxley, *Darwiniana* (New York: D. Appleton, 1908 [1893]).

criticism. Instead of abusive nonsense, which merely discredited its writers, we read essays, which are, at worst, more or less intelligent and appreciative; while, sometimes, like that which appeared in the *North British Review* for 1867, they have a real and permanent value.

The several publications of Mr. Wallace and Mr. Mivart contain discussions of some of Mr. Darwin's views, which are worthy of particular attention, not only on account of the acknowledged scientific competence of these writers, but because they exhibit an attention to those philosophical questions which underlie all physical science, which is as rare as it is needful. And the same may be said of an article in the *Quarterly Review* for July 1871, the comparison of which with an article in the same [*Quarterly*] *Review* for July 1860, is perhaps the best evidence which can be brought forward of the change which has taken place in public opinion on "Darwinism."

The *Quarterly* reviewer admits "the certainty of the action of natural selection" (p. 49); and further allows that there is an *a priori* probability in favor of the evolution of man from some lower animal form, if these lower animal forms themselves have arisen by evolution.

Mr. Wallace and Mr. Mivart go much further than this. They are as stout believers in evolution as Mr. Darwin himself; but Mr. Wallace denies that man can have been evolved from a lower animal by that process of natural selection which he, with Mr. Darwin, holds to have been sufficient for the evolution of all animals below man; while Mr. Mivart, admitting that natural selection has been one of the conditions of the evolution of the animals below man, maintains that natural selection must, even in their case, have been supplemented by "some other cause"—of the nature of which, unfortunately, he does not give us any idea. Thus Mr. Mivart is less of a Darwinian than Mr. Wallace, for he has less faith in the power of natural selection. But he is more of an evolutionist than Mr. Wallace, because Mr. Wallace thinks it necessary to call in an intelligent agent—a sort of supernatural Sir John Sebright—to produce even the animal frame of man; while Mr. Mivart requires no Divine assistance till he comes to man's soul.

The *Quarterly* reviewer and Mr. Mivart base their objections to the evolution of the mental faculties of man from those of some lower animal form upon what they maintain to be a difference in kind between the mental and moral faculties of men and brutes; and I have endeavored to show, by exposing the utter unsoundness of their philosophical basis, that these objections are devoid of importance.

The objections which Mr. Wallace brings forward to the doctrine of the evolution of the mental faculties of man from those of brutes by natural causes, are of a different order, and require separate consideration.

If I understand him rightly, he by no means doubts that both the bodily and the mental faculties of man have been evolved from those of some lower animal; but he is of opinion that some agency beyond that which has been concerned in the evolution of ordinary animals has been operative in the case of man. "A superior intelligence has guided the development of man in a definite direction and for a special purpose, just as man guides the development of many animal and vegetable forms." I understand this to mean that, just as the rock-pigeon has been produced by natural causes, while the evolution of the tumbler from the blue rock has required the special intervention of the intelligence of man, so some anthropoid form may have been evolved by variation and natural selection; but it could never have given rise to man, unless some superior intelligence had played the part of the pigeon-fancier.

According to Mr. Wallace, "whether we compare the savage with the higher developments of man, or with the brutes around him, we are alike driven to the conclusion, that, in his large and well-developed brain, he possesses an organ quite disproportioned to his requirements" (p. 343); and he asks, "What is there in the life of the savage but the satisfying of the cravings of appetite in the simplest and easiest way? What thoughts, idea, or actions are there that raise him many grades above the elephant or the ape?" (p. 342). I answer Mr. Wallace by citing a remarkable passage which occurs in his instructive paper on "Instinct in Man and Animals."

> Savages make long journeys in many directions, and, their whole faculties being directed to the subject, they gain a wide and accurate knowledge of the topography, not only of their own district, but of all the regions round about. Everyone who has travelled in a new direction communicates his knowledge to those who have travelled less, and descriptions of routes and localities, and minute incidents of travel, form one of the main staples of conversation around the evening fire. Every wanderer or captive from another tribe adds to the store of information, and, as the very existence of individuals and of whole families and tribes depends upon the completeness of this knowledge, all the acute perceptive facul-

ties of the adult savage are directed to acquiring and perfecting it. The good hunter or warrior thus comes to know the bearing of every hill and mountain range, the directions and junctions of all the streams, the situation of each tract characterized by peculiar vegetation, not only within the area he has himself traversed, but perhaps for a hundred miles around it. His acute observation enables him to detect the slightest undulations of the surface, the various changes of subsoil and alterations in the character of the vegetation that would be quite imperceptible to a stranger. His eye is always open to the direction in which he is going; the mossy side of trees, the presence of certain plants under the shade of rocks, the morning and evening flight of birds, are to him indications of direction almost as sure as the sun in the heavens (pp. 207, 208).

I have seen enough of savages to be able to declare that nothing can be more admirable than this description of what a savage has to learn. But it is incomplete. Add to all this the knowledge which a savage is obliged to gain of the properties of plants, of the characters and habits of animals, and of the minute indications by which their course is discoverable: consider that even an Australian can make excellent baskets and nets, and neatly fitted and beautifully balanced spears; that he learns to use these so as to be able to transfix a quartern loaf at sixty yards; and that very often, as in the case of the American Indians, the language of a savage exhibits complexities which a well-trained European finds it difficult to master: consider that every time a savage tracks his game he employs a minuteness of observation, and an accuracy of inductive and deductive reasoning which, applied to other matters, would assure some reputation to a man of science, and I think we need ask no further why he possesses such a fair supply of brains. In complexity and difficulty, I should say that the intellectual labor of a "good hunter or warrior" considerably exceeds that of an ordinary Englishman. The Civil Service Examiners are held in great terror by young Englishmen; but even their ferocity never tempted them to require a candidate to possess such a knowledge of a parish as Mr. Wallace justly points out savages may possess of an area a hundred miles or more in diameter.

But suppose, for the sake of argument, that a savage has more brains than seems proportioned to his wants, all that can be said is that the objection to natural selection, if it be one, applies quite as strongly to the lower animals. The brain of a porpoise is quite wonderful for its mass, and for the development of the cerebral convolutions. And yet since we have ceased to credit the story of Arion, it is hard to believe that porpoises are much troubled with intellect; and still more difficult is it to imagine that their big brains are only a preparation for the advent of some accomplished cetacean of the future. Surely, again, a wolf must have too much brains, or else how is it that a dog with only the same quantity and form of brain is able to develop such singular intelligence? The wolf stands to the dog in the same relation as the savage to the man; and, therefore, if Mr. Wallace's doctrine holds good, a higher power must have superintended the breeding up of wolves from some inferior stock, in order to prepare them to become dogs.

Mr. Wallace further maintains that the origin of some of man's mental faculties by the preservation of useful variations is not possible. Such, for example, are "the capacity to form ideal conceptions of space and time, of eternity and infinity; the capacity for intense artistic feelings of pleasure in form, color, and composition; and for those abstract notions of form and number which render geometry and arithmetic possible." "How," he asks, "were all or any of these faculties first developed, when they could have been of no possible use to man in his early stages of barbarism?"

Surely the answer is not far to seek. The lowest savages are as devoid of any such conceptions as the brutes themselves. What sort of conceptions of space and time, of form and number, can be possessed by a savage who has not got so far as to be able to count beyond five or six, who does not know how to draw a triangle or a circle, and has not the remotest notion of separating the particular quality we call form, from the other qualities of bodies? None of these capacities are exhibited by men, unless they form part of a tolerably advanced society. And, in such a society, there are abundant conditions by which a selective influence is exerted in favor of those persons who exhibit an approximation towards the possession of these capacities.

The savage who can amuse his fellows by telling a good story over the nightly fire, is held by them in esteem and rewarded, in one way or another, for so doing—in other words, it is an advantage to him to possess this power. He who can carve a paddle, or the figurehead of a canoe better, similarly profits beyond his duller neighbor. He who counts a little better than others, gets most yams when barter is going on, and forms the shrewdest estimate of the numbers of an opposing tribe. The experience of daily life shows that the conditions of our present social existence exercise the most extraordinarily powerful selective influence in favor of novelists, artists, and strong intellects of all kinds; and it seems unquestionable that all forms of social existence must have had the same tendency, if we consider the indisputable facts that even animals possess the power of distinguishing form and number, and that they are capable of deriving pleasure from particular forms and sounds. If we admit, as Mr. Wallace does, that the lowest savages are not raised "many grades above the elephant and the ape"; and if we further admit, as I contend must be admitted, that the conditions of social life tend, powerfully, to give an advantage to those individuals who vary in the direction of intellectual or aesthetic excellence, what is there to interfere with the belief that these higher faculties, like the rest, owe their development to natural selection?

Finally, with respect to the development of the moral sense out of the simple feelings of pleasure and pain, liking and disliking, with which the lower animals are provided, I can find nothing in Mr. Wallace's reasonings which has not already been met by Mr. Mill, Mr. Spencer, or Mr. Darwin.

MAN AND THE LOWER ANIMALS

Thus, whatever system of organs be studied, the comparison of their modifications in the ape series leads to one and the same result—that the structural differences which separate Man from the Gorilla and the Chimpanzee are not so great as those which separate the Gorilla from the lower apes.

And thus the sagacious foresight of the great lawgiver of systematic zoology, Linnaeus, becomes justified, and a century of anatomical research brings us back to his conclusion, that man is a member of the same order (for which the Linnaean term *Primates* ought to be retained) as the Apes and Lemurs. This order is now divisible into seven families, of about equal systematic value.

But if Man be separated by no greater structural barrier from the brutes than they are from one another—then it seems to follow that

Abridged from T. H. Huxley, *Man's Place in Nature* (London: Macmillan, 1906 [1863]).

if any process of physical causation can be discovered by which the genera and families of ordinary animals have been produced, that process of causation is amply sufficient to account for the origin of Man. In other words, if it could be shown that the Marmosets, for example, have arisen by gradual modification of the ordinary Platyrhini, or that both Marmosets and Platyrhini are modified ramifications of a primitive stock—then, there would be no rational ground for doubting that man might have originated, in the one case, by the gradual modification of a man-like ape; or, in the other case, as a ramification of the same primitive stock as those apes.

At the present moment, but one such process of physical causation has any evidence in its favor; or, in other words, there is but one hypothesis regarding the origin of species of animals in general which has any scientific existence—that propounded by Mr. Darwin.

In view of the intimate relations between Man and the rest of the living world, and between the forces exerted by the latter and all other forces, I can see no excuse for doubting that all are coordinated terms of Nature's great progression, from the formless to the formed —from the inorganic to the organic—from blind force to conscious intellect and will.

Science has fulfilled her function when she has ascertained and enunciated truth; and were these pages addressed to men of science only, I should now close this essay, knowing that my colleagues have learned to respect nothing but evidence, and to believe that their highest duty lies in submitting to it, however it may jar against their inclinations.

But desiring, as I do, to reach the wider circle of the intelligent public, it would be unworthy cowardice were I to ignore the repugnance with which the majority of my readers are likely to meet the conclusions to which the most careful and conscientious study I have been able to give to this matter, has led me.

On all sides I shall hear the cry—"We are men and women, not a mere better sort of apes, a little longer in the leg, more compact in the foot, and bigger in brain than your brutal Chimpanzees and Gorillas. The power of knowledge—the conscience of good and evil —the pitiful tenderness of human affections, raise us out of all real fellowship with the brutes, however closely they may seem to approximate us."

To this I can only reply that the exclamation would be most just and would have my own entire sympathy, if it were only relevant. But, it is not I who seek to base Man's dignity upon his great toe, or insinuate that we are lost if an Ape has a hippocampus minor. On the contrary, I have done my best to sweep away this vanity. I have endeavored to show that no absolute structural line of demarcation, wider than that between the animals which immediately succeed us in the scale, can be drawn between the animal world and ourselves; and I may add the expression of my belief that the attempt to draw a psychical distinction is equally futile, and that even the highest faculties of feeling and of intellect begin to germinate in lower forms of life. At the same time, no one is more strongly convinced than I am of the vastness of the gulf between civilized man and the brutes; or is more certain that whether *from* them or not, he is assuredly not *of* them. No one is less disposed to think lightly of the present dignity, or disparagingly of the future hopes, of the only consciously intelligent denizen of this world.

We are indeed told by those who assume authority in these matters, that the two sets of opinions are incompatible, and that the belief in the unity of origin of man and brutes involves the brutalization and degradation of the former. But is this really so? Could not a sensible child confute by obvious arguments, the shallow rhetoricians who would force this conclusion upon us? Is it, indeed, true, that the poet, or the philosopher, or the artist whose genius is the glory of his age, is degraded from his high estate by the undoubted historical probability, not to say certainty, that he is the direct descendant of some naked and bestial savage, whose intelligence was just sufficient to make him a little more cunning than the Fox, and by so much more dangerous than the Tiger? Or is he bound to howl and grovel on all fours because of the wholly unquestionable fact, that he was once an egg, which no ordinary power of discrimination could distinguish from that of a Dog? Or is the philanthropist, or the saint, to give up his endeavors to lead a noble life, because the simplest study of man's nature reveals, at its foundations, all the selfish passions, and fierce appetites of the merest quadruped? Is mother-love vile because a hen shows it, or fidelity base because dogs possess it?

Thoughtful men, once escaped from the blinding influences of traditional prejudice, will find in the lowly stock whence Man has sprung, the best evidence of the splendor of his capacities; and will discern in his long progress through the past, a reasonable ground of faith in his attainment of a nobler future.

They will remember that in comparing civilized man with the animal world, one is as the Alpine traveller, who sees the mountains soaring into the sky and can hardly discern where the deep shadowed crags and roseate peaks end, and where the clouds of heaven begin. Surely the awe-struck voyager may be excused if, at first, he refuses to believe the geologist, who tells him that these glorious masses are, after all, the hardened mud of primeval seas, or the cooled slag of subterranean furnaces—of one substance with the dullest clay, but raised by inward forces to that place of proud and seemingly inaccessible glory.

But the geologist is right; and due reflection on his teachings, instead of diminishing our reverence and our wonder, adds all the force of intellectual sublimity to the mere aesthetic intuition of the uninstructed beholder.

And after passion and prejudice have died away, the same result will attend the teachings of the naturalist respecting that great Alps and Andes of the living world—Man. Our reverence for the nobility of manhood will not be lessened by the knowledge that Man is, in substance and in structure, one with the brutes; for, he alone possesses the marvelous endowment of intelligible and rational speech, whereby, in the secular period of his existence, he has slowly accumulated and organized the experience which is almost wholly lost with the cessation of every individual life in other animals; so that, now, he stands raised upon it as on a mountain top, far above the level of his humble fellows, and transfigured from his grosser nature by reflecting, here and there, a ray from the infinite source of truth.

SCIENCE AND CHRISTIAN TRADITION

Much has yet to be learned, but, at present, natural knowledge affords no support to the notion that men have fallen from a higher

Abridged from T. H. Huxley, *Science and Christian Tradition* (London: Macmillan, 1895).

to a lower state. On the contrary, everything points to a slow natural evolution; which, favored by the surrounding conditions in such localities as the valleys of the Yang-tse-kang, the Euphrates, and the Nile, reached a relatively high pitch, five or six thousand years ago; while, in many other regions, the savage condition has persisted down to our day. In all this vast lapse of time there is not a trace of the occurrence of any general destruction of the human race; not the smallest indication that man has been treated on any other principles than the rest of the animal world.

That the doctrine of evolution implies a former state of innocence of mankind is quite true; but, as I have remarked, it is the innocence of the ape and of the tiger, whose acts, however they may run counter to the principles of morality, it would be absurd to blame. The lust of the one and the ferocity of the other are as much provided for in their organization, are as clear evidences of design, as any other features that can be named.

Observation and experiment upon the phenomena of society soon taught men that, in order to obtain the advantages of social existence, certain rules must be observed. Morality commenced with society. Society is possible only upon the condition that the members of it shall surrender more or less of their individual freedom of action. In primitive societies, individual selfishness is a centrifugal force of such intensity that it is constantly bringing the social organization to the verge of destruction. Hence the prominence of the positive rules of obedience to the elders; of standing by the family or the tribe in all emergencies; of fulfilling the religious rites, nonobservance of which is conceived to damage it with the supernatural powers, belief in whose existence is one of the earliest products of human thought; and of the negative rules, which restrain each from meddling with the life or property of another.

The highest conceivable form of human society is that in which the desire to do what is best for the whole, dominates and limits the action of every member of that society. The more complex the social organization the greater the number of acts from which each man must abstain, if he desires to do that which is best for all. Thus the progressive evolution of society means increasing restriction of individual freedom in certain directions.

With the advance of civilization, and the growth of cities and of nations by the coalescence of families and of tribes, the rules which constitute the common foundation of morality and of law became more numerous and complicated, and the temptations to break or evade many of them stronger. In the absence of a clear apprehension of the natural sanctions of these rules, a supernatural sanction was assumed; and imagination supplied the motives which reason was supposed to be incompetent to furnish. Religion, at first independent of morality, gradually took morality under its protection; and the supernaturalists have ever since tried to persuade mankind that the existence of ethics is bound up with that of supernaturalism.

I am not of that opinion. But, whether it is correct or otherwise, it is very clear to me that, as Beelzebub is not to be cast out by the aid of Beelzebub, so morality is not to be established by immorality. It is, we are told, the special peculiarity of the devil that he was a liar from the beginning. If we set out in life with pretending to know that which we do not know; with professing to accept for proof evidence which we are well aware is inadequate; with wilfully shutting our eyes and our ears to facts which militate against this or that comfortable hypothesis; we are assuredly doing our best to deserve the same character.

Many seem to think that, when it is admitted that the ancient literature, contained in our Bibles, has no more claim to infallibility than any other ancient literature; when it is proved that the Israelites and their Christian successors accepted a great many supernaturalistic theories and legends which have no better foundation than those of heathenism, nothing remains to be done but to throw the Bible aside as so much waste paper.

I have always opposed this opinion. It appears to me that if there is anybody more objectionable than the orthodox Bibliolater it is the heterodox Philistine, who can discover in a literature which, in some respects, has no superior, nothing but a subject for scoffing and an occasion for the display of his conceited ignorance of the debt he owes to former generations.

Twenty-two years ago I pleaded for the use of the Bible as an instrument of popular education, and I venture to repeat what I then said:

"Consider the great historical fact that, for three centuries, this book has been woven into the life of all that is best and noblest in

English history; that it has become the national Epic of Britain and is as familiar to gentle and simple, from John o' Groat's House to Land's End, as Dante and Tasso once were to the Italians; that it is written in the noblest and purest English and abounds in exquisite beauties of mere literary form; and, finally, that it forbids the veriest hind, who never left his village, to be ignorant of the existence of other countries and other civilizations and of a great past, stretching back to the furthest limits of the oldest nations in the world. By the study of what other book could children be so much humanized and made to feel that each figure in that vast historical procession fills, like themselves, but a momentary space in the interval between the eternities; and earns the blessings or the curses of all time, according to its effort to do good and hate evil, even as they also are earning their payment for their work?"

At the same time, I laid stress upon the necessity of placing such instruction in lay hands; in the hope and belief, that it would thus gradually accommodate itself to the coming changes of opinion; that the theology and the legend would drop more and more out of sight, while the perennially interesting historical, literary, and ethical contents would come more and more into view.

I may add yet another claim of the Bible to the respect and the attention of a democratic age. Throughout the history of the Western world, the scriptures, Jewish and Christian, have been the great instigators of revolt against the worst forms of clerical and political despotism. The Bible has been the Magna Charta of the poor and of the oppressed; down to modern times, no state has had a constitution in which the interests of the people are so largely taken into account, in which the duties, so much more than the privileges, of rulers are insisted upon, as that drawn up for Israel in Deuteronomy and in Leviticus; nowhere is the fundamental truth that the welfare of the state, in the long run, depends on the uprightness of the citizen so strongly laid down. Assuredly, the Bible talks no trash about the rights of man; but it insists on the equality of duties, on the liberty to bring about that righteousness which is somewhat different from struggling for "rights"; on the fraternity of taking thought for one's neighbor as for one's self.

The enormous influence which has thus been exerted by the Jewish and Christian Scriptures has had no necessary connection with cosmogonies, demonologies, and miraculous interferences. Their

FIGURE 7. The Swedish novelist August Strindberg rightly said that Edvard Munch's lithograph "The Cry" (1895) depicted mankind's "scream of terror in the presence of Nature flushed with anger. . . ." (*Courtesy, Museum of Fine Arts, Boston. William Francis Warden Fund*)

strength lies in their appeals, not to the reason, but to the ethical sense. I do not say that even the highest biblical ideal is exclusive of others or needs no supplement. But I do believe that the human race is not yet, possibly may never be, in a position to dispense with it.

Thomas Hardy and Algernon Charles Swinburne
MANKIND'S POST-DARWINIAN STATUS: LONELINESS OR LIBERATION?

The poems of Thomas Hardy and Algernon Charles Swinburne (1837–1909) give two pictures of man's status in a Darwinian world. "Hap" captures Hardy's mood at the time that he lost his romantic and orthodox Christian faith. Far better would it be, he said, to become the object of a conscious scorn by a known, but vengeful God, than be the subject of chance forces in a world that humans can neither understand nor control. Believing that the way to "the Better" called for a "full look at the Worst," Hardy depicted humans in his novels and poems as struggling against the forces of nature, chance, and personal suffering and impulsiveness. Through "A Plaint to Man" he offered one solution to a "better" way, namely an honest recognition that aid will not come from gods—who are human creations—but through loving-kindness shared with friends.

Swinburne's response to Darwin was far closer to T. H. Huxley and Leslie Stephen than to Hardy. Believing that mankind had become a slave to the moral limitations and conventions of Christianity, Swinburne glorified in a new, liberating naturalism. Humans, gods, all things are products of Hertha (the personified goddess of earth and growth), and with God slain by modern thought, man has new life and freedom. The poetry of Alfred Lord Tennyson given in Section II offered a third, more traditional, but still revolutionary image of mankind's post-Darwinian status.

Hap

*If but some vengeful god would call to me
From up the sky, and laugh: "Thou suffering thing,
Know that thy sorrow is my ecstasy,
That thy love's loss is my hate's profiting!"*

*Then would I bear it, clench myself, and die,
Steeled by the sense of ire unmerited;
Half-eased in that a Powerfuller than I
Had willed and meted me the tears I shed.*

*But not so. How arrives it joy lies slain,
And why unblooms the best hope ever sown?*

"Hap" and "A Plaint of Man" (lines 1–6, 19–32) are from Thomas Hardy, *Collected Poems* (New York: Macmillan, 1926), pp. 7 and 306. Printed by the permission of the Trustees of the Hardy Estate; The Macmillan Company of Canada Limited; and Macmillan, London and Basingstoke.

"Hertha" selections from Algernon Charles Swinburne, *Selected Poems,* ed. by William Morton Payne (Boston: D. C. Heath, 1905).

—Crass Casualty obstructs the sun and rain,
And dicing Time for gladness casts a moan. . . .
These purblind Doomsters had as readily strown
Blisses about my pilgrimage as pain.

A Plaint to Man[1]

When you slowly emerged from the den of Time,
And gained percipience as you grew,
And fleshed you fair out of shapeless slime,

Wherefore, O Man, did there come to you
The unhappy need of creating me—
A form like your own—for praying to?

.   .   .

—But since I was framed in your first despair
The doing without me has had no play
In the minds of men when shadows scare;

And now that I dwindle day by day
Beneath the deicide eyes of seers
In a light that will not let me stay,

And to-morrow the whole of me disappears,
The truth should be told, and the fact be faced
That had best been faced in earlier years:

The fact of life with dependence placed
On the human heart's resource alone,
In brotherhood bonded close and graced

With loving-kindness fully blown,
And visioned help unsought, unknown.

Hertha[2]

I am that which began;
  Out of me the years roll;
Out of me God and man;
  I am equal and Whole;
God changes, and man, and the form of them bodily; I am the soul.

Before ever land was,
  Before ever the sea,

---

[1] Lines 1–6; 19–32.
[2] Lines 1–18, 71–76, 96–107, 156–167, 171–200.

Mankind's Post-Darwinian Status

>   Or soft hair of the grass,
>   Or fair limbs of the tree,
> Or the flesh-coloured fruit of my branches, I was, and thy soul was in me.
>
>   First life on my sources
>   First drifted and swam;
>   Out of me are the forces
>   That save it or damn;
> Out of me man and woman, and wild-beast and bird; before God was, I am.
>
> . . .
>
>   A creed is a rod,
>   And a crown is of night;
>   But this thing is God,
>   To be man with thy might,
> To grow straight in the strength of thy spirit, and live out thy life as the light.
>
> . . .
>
>   The tree many-rooted
>   That swells to the sky
>   With frondage red-fruited,
>   The life-tree am I;
> In the buds of your lives is the sap of my leaves: ye shall live and not die.
>
>   But the Gods of your fashion
>   That take and that give,
>   In their pity and passion
>   That scourge and forgive,
> They are worms that are bred in the bark that falls off: they shall die and not live.
>
> . . .
>
>   I bid you but be;
>   I have need not of prayer;
>   I have need of you free
>   As your mouths of mine air;
> That my heart may be greater within me, beholding the fruits of me fair.
>
>   O my sons, O too dutiful
>   Toward Gods not of me,
>   Was not I enough beautiful?
>   Was it hard to be free?
> For behold, I am with you, am in you and of you; look forth now and see.
>
> . . .
>
>   Lo, winged with world's wonders,
>   With miracles shod,

>   With the fires of his thunders
>   For raiment and rod,
> God trembles in heaven, and his angels are white with the terror of God.
>
>   For his twilight is come on him,
>   His anguish is here;
>   And his spirits gaze dumb on him,
>   Grown grey from his fear;
> And his hour taketh hold on him stricken, the last of his infinite year.
>
>   Thought made him and breaks him,
>   Truth slays and forgives;
>   But to you, as time takes him,
>   This new thing it gives,
> Even love, the beloved Republic, that feeds upon freedom and lives.
>
>   For truth only is living,
>   Truth only is whole,
>   And the love of his giving
>   Man's polestar and pole;
> Man, pulse of my centre, and fruit of my body, and seed of my soul.
>
>   One birth of my bosom;
>   One beam of mine eye;
>   One topmost blossom
>   That scales the sky;
> Man, equal and one with me, man that is made of me, man that is I.

Herbert Spencer
SOCIETY CONDITIONED BY EVOLUTION

On the continent, as well as in England and America, Herbert Spencer (1820–1903) was avidly received as the great social philosopher who applied the natural "laws" of science to the development of human ethics and society. Through numerous, incredibly clumsy, dry and pompous books and articles, Spencer popularized sociology as the objective study of society free from metaphysical and moral bias. In this reading Spencer summarized the indebtedness of sociology (as the "science" of human society and its development) to biology. He focuses on three issues: (1) how society on the organizational level is like a biological organism; (2) how via the struggle for survival with the consequent elimination of unfit social systems societies evolve naturally and inevitably like biological species; and (3) how these two "facts" demand laissez-faire social policies, often termed "Social Darwinism." With all social progress dependent on the natural laws of evolutionary development, government-sponsored social reforms lead to retrogression and disaster.

The parable of the sower has its application to the progress of science. Time after time new ideas are sown and do not germinate, or, having germinated, die for lack of fit environments, before they are at last sown under such conditions as to take root and flourish. Among other instances of this, one is supplied by the history of the truth here to be dwelt on—the dependence of sociology on biology.

That there is a real analogy between an individual organism and a social organism, becomes undeniable when certain necessities determining structure are seen to govern them in common.

Mutual dependence of parts is that which initiates and guides organization of every kind. So long as, in a mass of living matter, all parts are alike, and all parts similarly live and grow without aid from one another, there is no organization: the undifferentiated aggregate of protoplasm thus characterized, belongs to the lowest grade of living things. Without distinct faculties, and capable of but the feeblest movements, it cannot adjust itself to circumstances; and is at the mercy of environing destructive actions. The changes by which this structureless mass becomes a structured mass, having the characters and powers possessed by what we call an organism,

Abridged from Herbert Spencer, *The Study of Sociology* (New York: D. Appleton, 1882).

are changes through which its parts lose their original likenesses; and do this while assuming the unlike kinds of activity for which their respective positions toward one another and surrounding things fit them. These differences of function, and consequent differences of structure, at first feebly marked, slight in degree, and few in kind, become, as organization progresses, definite and numerous; and in proportion as they do this the requirements are better met. Now structural traits expressible in the same language, distinguish lower and higher types of societies from one another; and distinguish the earlier stages of each society from the later. Primitive tribes show no established contrasts of parts. At first all men carry on the same kinds of activities, with no dependence on one another, or but occasional dependence. There is not even a settled chieftainship; and only in times of war is there a spontaneous and temporary subordination to those who show themselves the best leaders. From the small unformed social aggregates thus characterized, the progress is toward social aggregates of increased size, the parts of which acquire unlikenesses that become ever greater, more definite, and more multitudinous. The units of the society as it evolves, fall into different orders of activities, determined by differences in their local conditions or their individual powers; and there slowly result permanent social structures, of which the primary ones become decided while they are being complicated by secondary ones, growing in their turns decided, and so on.

Similarly, development of the individual organism, be its class what it may, is always accompanied by development of a nervous system which renders the combined actions of the parts prompt and duly proportioned, so making possible the adjustments required for meeting the varying contingencies; while, along with development of the social organism, there always goes development of directive centers, general and local, with established arrangements for interchanging information and instigation, serving to adjust the rates and kinds of activities going on in different parts.

Now if there exists this fundamental kinship, there can be no rational apprehension of the truths of sociology until there has been reached a rational apprehension of the truths of biology. The services of the two sciences are, indeed, reciprocal. We have but to glance back at its progress, to see that biology owes the cardinal idea on which we have been dwelling, to sociology; and that having derived from sociology this explanation of development, it gives it back to sociology greatly increased in definiteness, enriched by countless illustrations, and fit for extension in new directions.

* * *

Turn we now from the indirect influence which biology exerts on sociology, by supplying it with rational conceptions of social development and organization, to the direct influence it exerts by furnishing an adequate theory of the social unit—Man. For while biology is mediately connected with sociology by a certain parallelism between the groups of phenomena they deal with, it is immediately connected with sociology by having within its limits this creature whose properties originate social evolution. The human being is at once the terminal problem of biology and the initial factor of sociology.

If Man were uniform and unchangeable, so that those attributes of him which lead to social phenomena could be learnt and dealt with as constant, it would not much concern the sociologist to make himself master of other biological truths than those cardinal ones above dwelt upon. But since, in common with every other creature, Man is modifiable—since his modifications, like those of every other creature, are ultimately determined by surrounding conditions—and since surrounding conditions are in part constituted by social arrangements; it becomes requisite that the sociologist should acquaint himself with the laws of modification to which organized beings in general conform. Unless he does this he must continually err, both in thought and deed. As thinker, he will fail to understand the increasing action and reaction of institutions and character, each slowly modifying the other through successive generations. As actor, his furtherance of this or that public policy, being unguided by a true theory of the effects wrought on citizens, will probably be mischievous rather than beneficial; since there are more ways of going wrong than of going right. How needful is enlightenment on this point, will be seen on remembering that scarcely anywhere is attention given to the modifications which a new agency, political or other, will produce in men's natures. Immediate influence on actions is alone contemplated; and the immeasurably more important influence on the bodies and minds of future generations, is wholly ignored.

Yet the biological truths which should check this random political speculation and rash political action, are conspicuous; and might, one would have thought, have been recognized by everyone, even without special preparation in biology. That faculties and powers of all orders, while they grow by exercise, dwindle when not used; and that alterations of nature descend to posterity; are facts continually thrust on men's attention, and more or less admitted by each.

To give a more definite and effective shape to this general inference, let me here comment on certain courses pursued by philanthropists and legislators eager for immediate good results, but pursued without regard to biological truths which, if borne in mind, would make them hesitate if not desist.

Every species of creature goes on multiplying till it reaches the limit at which its mortality from all causes balances its fertility. Diminish its mortality by removing or mitigating any one of these causes, and inevitably its numbers increase until mortality and fertility are again in equilibrium. However many injurious influences are taken away, the same thing holds; for the reason that the remaining injurious influences grow more intense. Either the pressure on the means of subsistence becomes greater; or some enemy of the species, multiplying in proportion to the abundance of its prey, becomes more destructive; or some disease, encouraged by greater proximity, becomes more prevalent. This general truth, everywhere exemplified among inferior races of beings, holds of the human race.

Let us here glance at the relation between this general truth and the legislative measures adopted to ward off certain causes of death. Every individual eventually dies from inability to withstand some environing action. It may be a mechanical force that cannot be resisted by the strengths of his bodily structures; it may be a deleterious gas which, absorbed into his blood, so deranges the processes throughout his body as finally to overthrow their balance; or it may be an absorption of his bodily heat by surrounding things, that is too great for his enfeebled functions to meet. In all cases, however, it is one, or some, of the many forces to which he is exposed, and in presence of which his vital activities have to be carried on. He may succumb early or late, according to the goodness of his structure and the incidents of his career. But in the natural working of things, those having imperfect structures succumb before they have offspring: leaving those with fitter structures to produce the next generation. And obviously, the working of this process is such that as many will continue to live and to reproduce as can do so under the conditions then existing: if the assemblage of influences becomes more difficult to withstand, a larger number of the feebler disappear early; if the assemblage of influences is made more favorable by the removal of, or mitigation of, some unfavorable influence, there is an increase in the number of the feebler who survive and leave posterity. Hence two proximate results, conspiring to the same ultimate result. First, population increases at a greater rate than it would otherwise have done: so subjecting all persons to certain other destroying agencies in more-intense forms. Second, by intermarriage of the feebler who now survive, with the stronger who would otherwise have alone survived, the general constitution is brought down to the level of strength required to meet these more-favorable conditions. That is to say, there by-and-by arises a state of things under which a general decrease in the power of withstanding this mitigated destroying cause, and a general increase in the activity of other destroying causes, consequent on greater numbers, bring mortality and fertility into the same relation as before—there is a somewhat larger number of a somewhat weaker race.

Other evils, no less serious, are entailed by legislative actions and by actions of individuals, single and combined, which overlook or disregard a kindred biological truth. Besides an habitual neglect of the fact that the quality of a society is physically lowered by the artificial preservation of its feeblest members, there is an habitual neglect of the fact that the quality of a society is lowered morally and intellectually, by the artificial preservation of those who are least able to take care of themselves.

If anyone denies that children bear likenesses to their progenitors in character and capacity—if he holds that men whose parents and grandparents were habitual criminals, have tendencies as good as those of men whose parents and grandparents were industrious and upright, he may consistently hold that it matters not from what families in a society the successive generations descend. He may think it just as well if the most active, and capable, and prudent, and conscientious people die without issue; while many children are left by the reckless and dishonest. But whoever does not espouse so insane a proposition, must admit that social arrangements which retard the multiplication of the mentally best, and facilitate the multiplication of the mentally worst, must be extremely injurious.

For if the unworthy are helped to increase, by shielding them from that mortality which their unworthiness would naturally entail, the effect is to produce, generation after generation, a greater unworthiness. From diminished use of self-conserving faculties already deficient, there must result, in posterity, still smaller amounts of self-conserving faculties. The general law which we traced above in its bodily applications, may be traced here in its mental applications. Removal of certain difficulties and dangers which have to be met by intelligence and activity, is followed by a decreased ability to meet difficulties and dangers. Among children born to the more capable who marry with the less capable, thus artificially preserved, there is not simply a lower average power of self-preservation than would else have existed, but the incapacity reaches in some cases a greater extreme. Smaller difficulties and dangers become fatal in proportion as greater ones are warded off. Nor is this the whole mischief. For such members of a population as do not take care of themselves, but are taken care of by the rest, inevitably bring on the rest extra exertion; either in supplying them with the necessaries of life, or in maintaining over them the required supervision, or in both. That is to say, in addition to self-conservation and the conservation of their own offspring, the best, having to undertake the conservation of the worst, and of their offspring, are subject to an overdraw upon their energies. In some cases this stops them from marrying; in other cases it diminishes the numbers of their children; in other cases it causes inadequate feeding of their children; in other cases it brings their children to orphanhood—in every way tending to arrest the increase of the best, to deteriorate their constitutions, and to pull them down toward the level of the worst.

Fostering the good-for-nothing at the expense of the good, is an extreme cruelty. It is a deliberate storing-up of miseries for future generations. There is no greater curse to posterity than that of bequeathing them an increasing population of imbeciles and idlers and criminals. To aid the bad in multiplying, is, in effect, the same as maliciously providing for our descendants a multitude of enemies. It may be doubted whether the maudlin philanthropy which, looking only at direct mitigations, persistently ignores indirect mischiefs, does not inflict a greater total of misery than the extremest selfishness inflicts. Refusing to consider the remote influences of his incontinent generosity, the thoughtless giver stands but a degree above the drunkard who thinks only of today's pleasure and ignores tomorrow's pain, or the spendthrift who seeks immediate delights at the cost of ultimate poverty. In one respect, indeed, he is worse; since, while getting the present pleasure produced in giving pleasure, he leaves the future miseries to be borne by others—escaping them himself. And calling for still stronger reprobation is that scattering of money prompted by misinterpretation of the saying that "charity covers a multitude of sins." For in the many whom this misinterpretation leads to believe that by large donations they can compound for evil deeds, we may trace an element of positive baseness—an effort to get a good place in another world, no matter at what injury to fellow-creatures.

How far the mentally superior may, with a balance of benefit to society, shield the mentally inferior from the evil results of their inferiority, is a question too involved to be here discussed at length. Doubtless it is in the order of things that parental affection, the regard of relatives, and the spontaneous sympathy of friends and even of strangers, should mitigate the pains which incapacity has to bear, and the penalties which unfit impulses bring round. Doubtless, in many cases the reactive influence of this sympathetic care which the better take of the worse, is morally beneficial, and in a degree compensates by good in one direction for evil in another. It may be fully admitted that individual altruism, left to itself, will work advantageously—wherever, at least, it does not go to the extent of helping the unworthy to multiply. But an unquestionable injury is done by agencies which undertake in a wholesale way to foster good-for-nothings: putting a stop to that natural process of elimination by which society continually purifies itself. For not only by such agencies is this preservation of the worst and destruction of the best carried further than it would else be, but there is scarcely any of that compensating advantage which individual altruism implies. A mechanically working state-apparatus, distributing money drawn from grumbling ratepayers, produces little or no moralizing effect on the capables to make up for multiplication of the incapables. Here, however, it is needless to dwell on the perplexing questions hence arising. My purpose is simply to show that a rational policy must recognize certain general truths of biology; and to insist that only when study of these general truths, as illustrated throughout the living world, has woven them into the conceptions of things, is there gained a strong conviction that disregard of them must cause enormous mischiefs.

* * *

Enough has been said in proof of that which was to be shown—the use of biological study as a preparation for grasping sociological truths.

The effect to be looked for from it, is that of giving strength and clearness to convictions otherwise feeble and vague. Sundry of the doctrines I have presented under their biological aspects, are doctrines admitted in considerable degrees. Such acquaintance with the laws of life as they have gathered incidentally, lead many to suspect that appliances for preserving the physically feeble, bring results that are not wholly good. Others there are who occasionally get glimpses of evils caused by fostering the reckless and the stupid. But their suspicions and qualms fail to determine their conduct, because the *inevitableness* of the bad consequences has not been made adequately clear by the study of biology at large. When countless illustrations have shown them that all strength, all faculty, all fitness, presented by every living thing, has arisen partly by a growth of each power consequent on exercise of it, and partly by the more frequent survival and greater multiplication of the better-endowed individuals, entailing gradual disappearance of the worse-endowed—when it is seen that all perfection, bodily and mental, has been achieved through this process, and that suspension of it must cause cessation of progress, while reversal of it would bring universal decay—when it is seen that the mischiefs entailed by disregard of these truths, though they may be slow, are certain; there comes a conviction that social policy must be conformed to them, and that to ignore them is madness.

The theory of progress disclosed by the study of sociology as science, is one which greatly moderates the hopes and the fears of extreme parties. . . . Evidently, so far as a doctrine can influence general conduct (which it can do, however, in but a comparatively small degree), the doctrine of evolution, in its social applications, is calculated to produce a *steadying* effect, alike on thought and action.

If, as seems likely, some should propose to draw the seemingly awkward corollary that it matters not what we believe or what we teach, since the process of social evolution will take its own course in spite of us; I reply that while this corollary is in one sense true, it is in another sense untrue. Doubtless, from all that has been said it follows that, supposing surrounding conditions continue the same, the evolution of a society cannot be in any essential way diverted from its general course; though it also follows (and here the corollary is at fault) that the thoughts and actions of individuals, being natural factors that arise in the course of the evolution itself, and aid in further advancing it, cannot be dispensed with, but must be severally valued as increments of the aggregate force producing change. But while the corollary is even here partially misleading, it is, in another direction, far more seriously misleading. For though the process of social evolution is in its general character so far predetermined . . . yet it is quite possible to perturb, to retard, or to disorder the process. The analogy of individual development again serves us. The unfolding of an organism after its special type, has its approximately uniform course taking its tolerably definite time; and no treatment that may be devised will fundamentally change or greatly accelerate these: the best that can be done is to maintain the required favorable conditions. But it is quite easy to adopt a treatment which shall dwarf, or deform, or otherwise injure: the processes of growth and development may be, and very often are, hindered or deranged, though they cannot be artificially bettered. Similarly with the social organism. Though, by maintaining favorable conditions, there cannot be more good than that of letting social progress go on unhindered; yet an immensity of mischief may be done in the way of disturbing and distorting and repressing, by policies carried out in pursuance of erroneous conceptions. And thus, notwithstanding first appearances to the contrary, there is a very important part to be played by a true theory of social phenomena.

A few words to those who think these general conclusions discouraging, may be added. Probably the more enthusiastic, hopeful of great ameliorations in the state of mankind, to be brought about rapidly by propagating this belief or initiating that reform, will feel that a doctrine negativing their sanguine anticipations takes away much of the stimulus to exertion. If large advances in human welfare can come only in the slow process of things, which will inevitably bring them; why should we trouble ourselves?

Doubtless it is true that on visionary hopes, rational criticisms have a depressing influence. It is better to recognize the truth, however. As between infancy and maturity there is no shortcut by which there may be avoided the tedious process of growth and development through insensible increments; so there is no way from the lower forms of social life to the higher, but one passing through small successive modifications. If we contemplate the order of nature, we see that everywhere vast results are brought about by accumulations of minute actions. The surface of the Earth has been sculptured by forces which in the course of a year produce alterations scarcely anywhere visible. Its multitudes of different organic forms have arisen by processes so slow, that, during the periods our observations extend over, the results are in most cases inappreciable. We must be content to recognize these truths and conform our hopes to them.

Thus, admitting that for the fanatic [few] wild anticipation is needful as a stimulus, and recognizing the usefulness of his delusion as adapted to his particular nature and his particular function, the man of higher type must be content with greatly moderated expectations, while he perseveres with undiminished efforts. He has to see how comparatively little can be done, and yet to find it worthwhile to do that little: so uniting philanthropic energy with philosophic calm.

T. H. Huxley
SOCIETY MODIFIED ACCORDING TO HUMAN STANDARDS

Throughout 1892 T. H. Huxley researched, wrote, and polished his famous Romanes Lecture of 1893 from which this excerpt is taken. In answering the question concerning the relationship between biological evolution and social morality, he opposed philosophers and ethicists like Herbert Spencer who believed that the "cosmic process" of evolution furnished normative ethical and social standards. Using Stoicism as one of his beginning points, Huxley argued that the posture of "living according to nature" did not and should not entail a resignation to the principles of competition and the survival of the fittest. Huxley's humanism, optimism and literary skills are displayed in his combating the "false analogies" and conclusions of laissez-faire Spencerians.

Man, the animal . . . has worked his way to the headship of the sentient world, and has become the superb animal which he is, in virtue of his success in the struggle for existence. The conditions having been of a certain order, man's organization has adjusted itself to them better than that of his competitors in the cosmic strife. In the case of mankind, the self-assertion, the unscrupulous seizing upon all that can be grasped, the tenacious holding of all that can be kept, which constitute the essence of the struggle for existence, have answered. For his successful progress, throughout the savage state, man has been largely indebted to those qualities which he shares with the ape and the tiger; his exceptional physical organization; his cunning, his sociability, his curiosity, and his imitativeness; his ruthless and ferocious destructiveness when his anger is roused by opposition.

But, in proportion as men have passed from anarchy to social organization, and in proportion as civilization has grown in worth, these deeply ingrained serviceable qualities have become defects. After the manner of successful persons, civilized man would gladly kick down the ladder by which he has climbed. He would be only too pleased to see "the ape and tiger die." But they decline to suit his convenience; and the unwelcome intrusion of these boon com-

Abridged from Thomas H. Huxley, *Evolution and Ethics* (London: Macmillan, 1895).

panions of his hot youth into the ranged existence of civil life adds pains and griefs, innumerable and immeasurably great, to those which the cosmic process necessarily brings on the mere animal. In fact, civilized man brands all these ape and tiger promptings with the name of sins; he punishes many of the acts which flow from them as crimes; and, in extreme cases, he does his best to put an end to the survival of the fittest of former days by axe and rope.

I have said that civilized man has reached this point; the assertion is perhaps too broad and general; I had better put it that ethical man has attained thereto. The science of ethics professes to furnish us with a reasoned rule of life; to tell us what is right action and why it is so. Whatever differences of opinion may exist among experts, there is a general consensus that the ape and tiger methods of the struggle for existence are not reconcilable with sound ethical principles.

The stoical summary of the whole duty of man, "Live according to nature," would seem to imply that the cosmic process is an exemplar for human conduct. Ethics would thus become applied Natural History. In fact, a confused employment of the maxim, in this sense, has done immeasurable mischief in later times. It has furnished an axiomatic foundation for the philosophy of philosophasters and for the moralizing of sentimentalists. But the Stoics were, at bottom, not merely noble, but sane, men; and if we look closely into what they really meant by this ill-used phrase, it will be found to present no justification for the mischievous conclusions that have been deduced from it.

In the language of the [Stoic], "nature" was a word of many meanings. There was the "nature" of the cosmos and the "nature" of man. In the latter, the animal "nature," which man shares with . . . the living part of the cosmos, was distinguished from a higher "nature." Even in this higher nature there were grades of rank. The logical faculty is an instrument which may be turned to account for any purpose. The passions and the emotions are so closely tied to the lower nature that they may be considered to be pathological, rather than normal, phenomena. The one supreme, hegemonic, faculty, which constitutes the essential "nature" of man, is most nearly represented by that which, in the language of a later philosophy, has been called the pure reason. It is this "nature" which

Society Modified According to Human Standards

holds up the ideal of the supreme good and demands absolute submission of the will to its behests. It is this which commands all men to love one another, to return good for evil, to regard one another as fellow-citizens of one great state. Indeed, seeing that the progress towards perfection of a civilized state, or polity, depends on the obedience of its members to these commands, the Stoics sometimes termed the pure reason the "political" nature. Unfortunately, the sense of the adjective has undergone so much modification, that the application of it to that which commands the sacrifice of self to the common good would now sound almost grotesque.

But what part is played by the theory of evolution in this view of ethics? So far as I can discern, the ethical system of the Stoics, which is essentially intuitive, and reverences the categorical imperative as strongly as that of any later moralists, might have been just what it was if they had held any other theory; whether that of special creation, on the one side, or that of the eternal existence of the present order, on the other. To the Stoic, the cosmos had no importance for the conscience, except in so far as he chose to think it a pedagogue to virtue.

* * *

To what extent modern progress in natural knowledge, and, more especially, the general outcome of that progress in the doctrine of evolution, is competent to help us in the great work of helping one another?

The propounders of what are called the "ethics of evolution," when the "evolution of ethics" would usually better express the object of their speculations, adduce a number of more or less interesting facts and more or less sound arguments, in favor of the origin of the moral sentiments, in the same way as other natural phenomena, by a process of evolution. I have little doubt, for my own part, that they are on the right track; but as the immoral sentiments have no less been evolved, there is, so far, as much natural sanction for the one as the other. The thief and the murderer follow nature just as much as the philanthropist. Cosmic evolution may teach us how the good and the evil tendencies of man may have come about; but, in itself, it is incompetent to furnish any better reason why what we call good is preferable to what we call evil than we had before. Some day, I doubt not, we shall arrive at an understanding of the evolution of the aesthetic faculty; but all the understanding in the world will neither increase nor diminish the force of the intuition that this is beautiful and that is ugly.

There is another fallacy which appears to me to pervade the so-called "ethics of evolution." It is the notion that because, on the whole, animals and plants have advanced in perfection of organization by means of the struggle for existence and the consequent "survival of the fittest"; therefore men in society, men as ethical beings, must look to the same process to help them towards perfection. I suspect that this fallacy has arisen out of the unfortunate ambiguity of the phrase "survival of the fittest." "Fittest" has a connotation of "best"; and about "best" there hangs a moral flavor. In cosmic nature, however, what is "fittest" depends upon the conditions. Long since, I ventured to point out that if our hemisphere were to cool again, the survival of the fittest might bring about, in the vegetable kingdom, a population of more and more stunted and humbler and humbler organisms, until the "fittest" that survived might be nothing but lichens, diatoms, and such microscopic organisms as those which give red snow its color; while, if it became hotter, the pleasant valleys of the Thames and Isis might be uninhabitable by any animated beings save those that flourish in a tropical jungle. They, as the fittest, the best adapted to the changed conditions, would survive.

Men in society are undoubtedly subject to the cosmic process. As among other animals, multiplication goes on without cessation, and involves severe competition for the means of support. The struggle for existence tends to eliminate those less fitted to adapt themselves to the circumstances of their existence. The strongest, the most self-assertive, tend to tread down the weaker. But the influence of the cosmic process on the evolution of society is the greater the more rudimentary its civilization. Social progress means a checking of the cosmic process at every step and the substitution for it of another, which may be called the ethical process; the end of which is not the survival of those who may happen to be the fittest, in respect of the whole of the conditions which obtain, but of those who are ethically the best.

As I have already urged, the practice of that which is ethically best—what we call goodness or virtue—involves a course of conduct which, in all respects, is opposed to that which leads to success in the cosmic struggle for existence. In place of ruthless self-assertion it demands self-restraint; in place of thrusting aside, or treading down, all competitors, it requires that the individual shall not merely respect, but shall help his fellows; its influence is directed, not so much to the survival of the fittest, as to the fitting of as many as possible to survive. It repudiates the gladiatorial theory of existence. It demands that each man who enters into the enjoyment of the advantages of a polity shall be mindful of his debt to those who have laboriously constructed it; and shall take heed that no act of his weakens the fabric in which he has been permitted to live. Laws and moral precepts are directed to the end of curbing the cosmic process and reminding the individual of his duty to the community, to the protection and influence of which he owes, if not existence itself, at least the life of something better than a brutal savage.

It is from neglect of these plain considerations that the fanatical individualism of our time attempts to apply the analogy of cosmic nature to society. Once more we have a misapplication of the stoical injunction to follow nature; the duties of the individual to the state are forgotten, and his tendencies to self-assertion are dignified by the name of rights. It is seriously debated whether the members of a community are justified in using their combined strength to constrain one of their number to contribute his share to the maintenance of it; or even to prevent him from doing his best to destroy it. The struggle for existence, which has done such admirable work in cosmic nature, must, it appears, be equally beneficent in the ethical sphere. Yet if that which I have insisted upon is true; if the cosmic process has no sort of relation to moral ends; if the imitation of it by man is inconsistent with the first principles of ethics; what becomes of this surprising theory?

Let us understand, once for all, that the ethical progress of society depends, not on imitating the cosmic process, still less in running away from it, but in combating it. It may seem an audacious proposal thus to pit the microcosm against the macrocosm and to set man to subdue nature to his higher ends; but I venture to think that the great intellectual difference between the ancient times with which we have been occupied and our day, lies in the solid foundation we have acquired for the hope that such an enterprise may meet with a certain measure of success.

The history of civilization details the steps by which men have succeeded in building up an artificial world within the cosmos. Fragile reed as he may be, man, as Pascal says, is a thinking reed: there lies within him a fund of energy, operating intelligently and so far akin to that which pervades the universe, that it is competent to influence and modify the cosmic process. In virtue of his intelligence, the dwarf bends the Titan to his will. In every family, in every polity that has been established, the cosmic process in man has been restrained and otherwise modified by law and custom; in surrounding nature, it has been similarly influenced by the art of the shepherd, the agriculturist, the artisan. As civilization has advanced, so has the extent of this interference increased; until the organized and highly developed sciences and arts of the present day have endowed man with a command over the course of non-human nature greater than that once attributed to the magicians. The most impressive, I might say startling, of these changes have been brought about in the course of the last two centuries; while a right comprehension of the process of life and of the means of influencing its manifestations is only just dawning upon us. We do not yet see our way beyond generalities; and we are befogged by the obtrusion of false analogies and crude anticipations. But astronomy, physics, chemistry, have all had to pass through similar phases, before they reached the stage at which their influence became an important factor in human affairs. Physiology, psychology, ethics, political science, must submit to the same ordeal. Yet it seems to me irrational to doubt that, at no distant period, they will work as great a revolution in the sphere of practice.

The theory of evolution encourages no millennial anticipations. If, for millions of years, our globe has taken the upward road, yet, sometime, the summit will be reached and the downward route will be commenced. The most daring imagination will hardly venture upon the suggestion that the power and the intelligence of man can ever arrest the procession of the great year.

Moreover, the cosmic nature born with us and, to a large extent, necessary for our maintenance, is the outcome of millions of years of severe training, and it would be folly to imagine that a few centuries will suffice to subdue its masterfulness to purely ethical ends. Ethical nature may count upon having to reckon with a tenacious and powerful enemy as long as the world lasts. But, on the other

Society Modified According to Human Standards

hand, I see no limit to the extent to which intelligence and will, guided by sound principles of investigation, and organized in common effort, may modify the conditions of existence, for a period longer than that now covered by history. And much may be done to change the nature of man himself. The intelligence which has converted the brother of the wolf into the faithful guardian of the flock ought to be able to do something towards curbing the instincts of savagery in civilized men.

But if we may permit ourselves a larger hope of abatement of the essential evil of the world than was possible to those who, in the infancy of exact knowledge, faced the problem of existence more than a score of centuries ago, I deem it an essential condition of the realization of that hope that we should cast aside the notion that the escape from pain and sorrow is the proper object of life.

We have long since emerged from the heroic childhood of our race, when good and evil could be met with the same "frolic welcome"; the attempts to escape from evil, whether Indian or Greek, have ended in flight from the battlefield; it remains to us to throw aside the youthful overconfidence and the no less youthful discouragement of nonage. We are grown men, and must play the man

strong in will
To strive, to seek, to find, and not to yield,

cherishing the good that falls in our way, and bearing the evil, in and around us, with stout hearts set on diminishing it. So far, we all may strive in one faith towards one hope:

It may be that the gulfs will wash us down,
It may be we shall touch the Happy Isles,

. . . but something ere the end,
Some work of noble note may yet be done.

Suggestions for Additional Reading

The great quantity of literature on Darwin and Darwinism attests to the continuing significance of the issues raised by evolution. Given the scope and volume of this literature, critical bibliographical essays are often highly useful. Among these is the excellent review by Bert James Loewenberg, "Darwin and Darwin Studies, 1959–1963," *History of Science* 4 (1965): 15–54. In this evaluation of literature on Darwin and his influence after the widely heralded centennial of *The Origin of Species* in 1959, Loewenberg in effect brings the reader up to date on the availability of manuscript and primary sources, discusses the accuracy and usefulness of recent biographies of Darwin, and makes an assessment of the scientific status of Darwin's research. Loewenberg's essay includes his own thoughtful observations about Darwin's historical significance, humanist temperament, and abilities as a historian of the natural world. Still useful also—particularly for its description of Darwin's waning influence after 1880, and his dramatic scientific rehabilitation after 1929, and for its pinpointing of Darwin's discoveries—is Donald Fleming's "The Centenary of *The Origin of Species,*" *Journal of the History of Ideas* 20 (1959): 437–446. For a critical evaluation of studies in the history of science in the nineteenth century, see the review by Everett Mendelsohn, "The Biological Sciences in the Nineteenth Century," *History of Science* 3 (1964): 37–59.

Other bibliographical sources are suggested in the footnotes and bibliographies of recent intellectual histories that place Darwin's ideas against the backdrop of developing Western thought. For the intellectual history of modern Europe, see particularly the recent work by Willson H. Coates, Hayden V. White, and J. Salwyn Schapiro, *The Emergence of Liberal Humanism,* II (New York: McGraw-Hill, 1970). Also useful for a depiction of Darwin's general place in history is Roland N. Stromberg, *An Intellectual History of Modern Europe* (New York: Meredith, 1966); and the older, but still noteworthy study by John Herman Randall, Jr., *The Making of the Modern Mind,* rev. ed. (New York: Houghton Mifflin, 1940). Effective surveys of American developments relating to the "naturalization" of man and the world are those by Merle Curti, *The Growth of American Thought,* 3rd ed. (New York: Harper and Row, 1964), part VI; Stow Persons, *American Minds* (New York: Henry Holt, 1958), part IV; and the more detailed study by Paul F. Boller, Jr., *American Thought in Transition* (Chicago: Rand McNally, 1969).

Several more narrowly focused background studies are of significant value in relating Darwin and Darwinism to their times. Leslie Stephen's two-volume *History of English Thought in the Eighteenth Century* (New York: G. P. Putnam's Sons, 1881 [1876]) still represents a remarkably critical account of English philosophy, theology and literature before the romantic period. Focusing on the multifaceted conceptions of nature from Alexander Pope to William Wordsworth is the classical and readable study by Basil Willey, *The Eighteenth Century Background* (Boston: Beacon Press, 1961 [1940]). The best primary source collection for religion and philosophy in the century before Darwin is John Martin Creed and John S. BoysSmith, *Religious Thought in the Eighteenth Century* (Cambridge: University Press, 1934), which includes readings from William Paley. Several editions of the poetry of major figures such as William Cowper and William Wordsworth are available. Striking portraits of Victorian life, feeling, and thought are found in Walter E. Houghton, *The Victorian Frame of Mind, 1830–1870* (New Haven: Yale University Press, 1957); in the two studies by George Malcolm Young, entitled *Victorian England* (New York: Oxford University Press, 1957), and *Victorian Essays* (London: Oxford University Press, 1962); and in the admirable essays by Basil Willey, *Nineteenth-Century Studies: Coleridge to Matthew Arnold* (London: Chatto and Windus, 1949), and by D. C. Somervell, *English Thought in the Nineteenth Century* (New York: David McKay, 1965 [1929]).

The history and impact of science is discussed in several studies. A clear, accurate account of the influence of geology on religion and social thought in England from 1790 to 1850 is found in Charles Coulton Gillispie, *Genesis and Geology* (New York: Harper and Brothers, 1951). Except for its failure to display Darwin's creative discoveries, *Forerunners of Darwin, 1745–1859,* ed. by Bentley Glass, Owsei Temkin and William L. Strauss, Jr. (Baltimore: Johns Hopkins Press, 1959), is an outstanding collection of essays on evolutionary thought in science, philosophy and literature in Europe up to the time of the publication of *The Origin of Species.* Thinkers such as Buffon, Kant, Herder and Lamarck are considered. Concentrating especially on the study of man and fossils (with illustrations) up to and including Darwin is the excellent study by John C. Greene, *The*

Death of Adam (Ames, Iowa: Iowa State University Press, 1959). And as an able anthropologist Loren Eiseley in *Darwin's Century* (Garden City: Doubleday, 1961 [1958]), discusses the concept of evolution and its influence in geology, biology, and anthropology from Linnaeus to Darwin to the end of the nineteenth century. Brief primary readings from pre-Darwinian scientists and from secondary readings that debate the significance of the "forerunners" of Darwin are found in the reader by Philip Appleman, ed., *Darwin* (New York: W. W. Norton, 1970), pp. 3–49. The original, influential essay by Thomas Malthus has been reprinted: *Population: The First Essay,* with a foreword by Kenneth E. Boulding (Ann Arbor: University of Michigan Press, 1959).

Most of Charles Darwin's primary writings, including *The Voyage of the Beagle, The Origin of Species* and *The Descent of Man,* are available in several modern editions. Valuable, recent editions of Darwin's journal and his notebooks on the transmutation of species are also available for in-depth research: *Darwin's Journal* and *Notebooks on Transmutation of Species,* ed., with notes, by Gavin de Beer, in the *Bulletin of the British Museum (Natural History),* Historical Series, ii (London, 1959 and 1960). Also especially noteworthy are the following: the new edition of Darwin, *The Autobiography of Charles Darwin,* ed. by Francis Darwin (New York: Dover Publications, 1958 [1892]); the unexpurgated edition of Darwin's autobiography, ed. by Nora Barlow, *The Autobiography of Charles Darwin, 1809–1882* (New York: Harcourt, Brace, 1958), in which Darwin's animus toward religion is heightened beyond the expurgated version of the *Autobiography* edited by Francis Darwin; a hard-to-use, but critical text that records the changes made by Darwin in the *Origin* during its six editions from 1859 until 1872, ed. by Morse Peckham, *The Origin of Species by Charles Darwin: A Variorum Text* (Philadelphia: University of Pennsylvania Press, 1959); and two valuable reprintings of the first edition of the *Origin:* Darwin, *On the Origin of Species,* ed. by C. D. Darlington, in *Thinking Library* (London, 1950), and ed. by Ernst Mayr (Cambridge: Harvard University Press, 1964). A fine general reader containing lengthy selections from Darwin's major scientific works is *The Darwin Reader,* ed. by Marston Bates and Philip S. Humphrey (New York: Charles Scribner's Sons, 1956).

Biographies and specialized studies of Darwin's life and thought are often fascinating avenues for further research. The concise study by Gavin de Beer, *Charles Darwin* (New York: Doubleday, 1963) is probably the best single biography, certainly the best in terms of

Suggestions for Additional Reading

using recent scholarship on Darwin and in assessing Darwin's contributions in terms of modern science. Much more controversial because of her rejection of natural selection is the admirably written biography of Darwin by Gertrude Himmelfarb, *Darwin and the Darwinian Revolution* (Garden City: Doubleday, 1962), who is conversant with the Victorian period and who provides the reader with helpful analyses concerning Darwin's impact on religion, morality and social thought. Not to be dismissed are the popular, illustrated studies of Darwin's life and the voyage of the *Beagle* by Julian Huxley and H. B. D. Kettlewell, *Charles Darwin and His World* (New York: Viking, 1965), and Alan Moorehead, *Darwin and the Beagle* (New York: Harper and Row, 1969). Among the critical essays concerning Darwin's thought, the following are particularly significant: Sydney Smith, "The Origin of 'The Origin'," *The Advancement of Science*, no. 64 (March 1960), pp. 391–401; the brilliant, controversial analysis of Darwin as a Victorian fleeing from pain, feeling, and religion, by Donald Fleming, "Charles Darwin, the Anaesthetic Man," *Victorian Studies* 4 (1941): 219–236; the splendid, enduring essay on Darwin as a creative writer and pioneer of modern thought by Stanley Edgar Hyman, *The Tangled Bank* (New York: Atheneum, 1962), pp. 9–79, 425–447; and the careful study concerning Darwin's scientific creativity by Peter Vorzimmer, "Darwin, Malthus, and the Theory of Natural Selection," *Journal of the History of Ideas* 30 (October 1969): 527–542.

Several methods of analysis have and can be used in order to assess the impact of Darwin and Darwinism. Representative biographical studies of notable Darwinians or anti-Darwinians are those by William Irving, *Apes, Angels and Victorians* (New York: McGraw-Hill, 1955), the latter half of which is devoted to T. H. Huxley; by Wilma George, *Biologist Philosopher: A Study of the Life and Writings of Alfred Russel Wallace* (London: Abelard-Schuman, 1964); by Jacob W. Gruber, *A Conscience in Conflict: the Life of St. George Jackson Mivart* (New York: Columbia University Press, 1960); and by Milton Berman, *John Fiske: The Evolution of a Popularizer* (Cambridge: Harvard University Press, 1961). Darwin's direct and often indirect influence (delineated by Morse Peckham in terms of Darwinism and "Darwinisticism") in literature is explained in the books by Leo Justin Henkin, *Darwinism and the English Novel, 1860–1910* (New York: Corporate Press, 1963 [1940]), Lionel Stevenson, *Darwin Among the Poets* (Chicago: University of Chicago Press, 1932), and the abbreviated readings in the collection by Philip Appleman, ed., *Darwin,* op. cit., pp. 573–625. For Darwinism and religion, see particularly Owen Chadwick, *The Victorian Church,* II (New York: Oxford University Press, 1970); Desmond Bowen, *The Idea of the Victorian Church* (Montreal: McGill University Press, 1968), chapter IV (whose intellectual history is far better than his social history); Richard Hofstadter, *Social Darwinism in American Thought, 1860–1915* (Boston: Beacon Press, 1955); Edward A. White, *Science and Religion in American Thought* (Stanford: Stanford University Press, 1952); Gail Kennedy, ed., *Evolution and Religion* (Boston: D. C. Heath, 1957); and Harold Y. Vanderpool, "Charles Darwin and Darwinism," in *Critical Issues in Modern Religion,* Roger A. Johnson et al. (Englewood Cliffs, New Jersey: Prentice-Hall, 1973), pp. 77–113. Darwin's impact on social thought—often indirect and clearly more significant in America than in England—is the central concern of Hofstadter's *Social Darwinism in American Thought,* op. cit., and is debated and considered in the readings from Walter Bagehot, Peter Kropotkin and others in Appleman, ed., *Darwin,* op. cit., pp. 489–570, and in the recent study of Herbert Spencer's social thought by J. D. Y. Peel, *Herbert Spencer* (London: Heinemann Educational Books, 1971). Several of Spencer's major writings are still being printed.

Numerous works continue to evaluate Darwin's status among contemporary scientists and intellectuals. Vindication of the first edition of *The Origin of Species* began with two studies that viewed evolution in terms of the combined result of natural selection and genetic mutation, the first by Ronald Fisher, *Genetic Theory of Natural Selection* (New York: Dover Publications, 1958 [first ed. in 1929]), and the second by Julian Huxley, *Evolution: The Modern Synthesis* (New York: Harper and Brothers, 1942). More recent evaluations are contained in several impressive studies, including S. A. Barnett, ed., *A Century of Darwin* (Cambridge: Harvard University Press, 1958); Chicago Centennial, *Evolution After Darwin,* vols. I–III (Chicago: University of Chicago Press, 1960), which includes essays on the evolution of life, species, man, society and religion; and the recent, exciting collection of articles from *Scientific American* by Joseph G. Jorgensen, ed., *Biology and Culture in Modern Perspective* (San Francisco: W. H. Freeman, 1972).